Packaging in Glass

Packaging in GLASS

Brian Moody

M.A., F.Inst.P., F.S.G.T., M.Inst.Pkg.

Design and Development Manager
United Glass Limited

Hutchinson Benham, London

Hutchinson Benham Limited
3 Fitzroy Square, London WI P 6JD

An imprint of the Hutchinson Group

London Melbourne Sydney Auckland
Wellington Johannesburg and agencies
throughout the world

First published 1963
Second impression 1964
Third impression 1970
Revised edition 1977
© United Glass 1963, 1977

Designed and produced by
Hutchinson Benham Limited

Printed in Great Britain by
The Anchor Press Ltd, and bound by
Wm Brendon & Son Ltd, both of
Tiptree, Essex

ISBN 0 09 129490 8

Contents

Machinery lists

8 *Machinery lists*

Packing

Preface

When the first edition of this book was published in 1963 it was intended to explain the basic scientific principles which lay behind bottle design, manufacture and usage. It is gratifying to find, fourteen years later, that many parts of the original text remain as valid now as they did then; this is perhaps just as well, because I know of people in many parts of the world who are using the book regularly as a guide and reference source on glass packaging.

The main change which has taken place during the past decade is that the glass makers and glass users have achieved a greater degree of control over their operations – so the scientific principles now fit the facts and achievements better than they did then. In other words, packaging in glass has become more of a science and less of a craft.

Nevertheless, there were many points of detail in the first edition which required updating, and some sections where there has been a quiet revolution. For example, the new science of surface treatment has enabled glass makers to produce containers with better strength and performance than ever before.

Then there is the new approach to bottle inspection, using automatic machines as well as human inspectors, together with the now universal use of statistical quality control techniques. (The words 'quality control' did not appear anywhere in the first edition!)

Yet another development is the greater degree of sophistication which is now possible in glass design, while the improvements which have taken place in glass forming mean that most of the bottles themselves have changed since 1963. Most of them are significantly lighter, and so relatively more economical; and generally their designs are better adjusted to suit the job they have to do, both on the filling line and afterwards (i.e. throughout the chain of distribution or in the hands of the consumer).

As far as the filling line itself is concerned, the most noticeable advance is in the speed at which bottles and jars can be handled. Potential through-puts of some kinds of filling line are now two or three times as great as they were fourteen years ago. This has called for higher standards of performance from all the equipment in the line, and most of the chapters dealing with the various kinds of bottling plant have been substantially revised.

Techniques of handling both empty and full bottles have also changed considerably with the advent of bulk palletising on the one hand and shrink packaging on the other.

Finally, changes in legislation affecting bottlers and packers have created a number of new requirements, while the effects of certain directives issued by the European Economic Community are beginning to make themselves felt in no uncertain manner. All these matters are fully discussed in this new edition.

The book is still intended primarily for packers who use glass containers or who would like to have advice on using them. Bottle makers, package designers and suppliers of packaging equipment should also find the work useful as a source of reference. Last, but not least, members of the public – the people who buy and use the end products – are becoming increasingly interested (and sometimes critical) about what goes on in packaging, so they too should find some interesting and useful information.

As before, the book aims to be suitable for use as a text book for students and newcomers to packaging who wish to learn about the subject as quickly and painlessly as possible. Its usefulness as a reference source is aided by an extensive set of appendices, a number of cross-references in the text, and some intentional repetition.

DEFINITIONS OF TERMS USED

Although there is a combined index and glossary at the end of the book, it is useful to begin here by defining a few of the basic words which have to be used repeatedly.

Bottle: Any type of sealable glass container. (In most of the text no subdivision is necessary into narrow-neck bottles and wide-mouth jars.)

Closure: Any type of cap, cork, plug, etc., used to seal the bottle.

Product: The material (liquid or solid) which is put into the bottle.

Customer or packer: The people who buy the empty bottles and fill them.

Purchaser, user or consumer: The person who buys or acquires the filled bottle and/or makes use of the contents. Usually the terms are synonymous and are used indiscriminately, but sometimes the right one must be selected (for example, with jars of baby food).

Case: A generic term for any type of outer box in which bottles are stored and transported. It may be made of wood, metal, plastic or fibreboard; but if of fibreboard it must be of the type which can be completely closed and sealed.

Tray: An open-top box for bottles, often with hand-holes at the ends.

AN OUTLINE OF THE BOOK

Chapters 1 to 5 deal with the design, manufacture and properties of glass containers, while Chapters 6 to 13 are concerned with packaging methods. At the end of each chapter dealing with packaging machinery, a classified list is given of the main types and suppliers of equipment generally available in the UK. Chapter 14 discusses a subject which is of vital interest to many people – the effect of glass containers on the environment. Lastly, Chapters 15 and 16 give information about the past and the present in the general field of packaging in glass, with a glimpse at what may possibly happen in the future.

This is a technical book, and some use has to be made of physics and chemistry, and of mathematical and engineering principles; but I have tried to keep the approach simple enough not to require a qualified scientist to understand it. Nevertheless it is intended that the technical sections shall be scientifically accurate and complete. In some cases, where it has been necessary to shorten or simplify a scientific argument, the simplifications have been stated.

Metric units are used throughout, and these are defined and explained in Appendix A.

ACKNOWLEDGMENTS

Many of the illustrations have been generously supplied by British packers and by suppliers of packaging machinery and materials. Separate acknowledgments have not been placed under individual photographs, but the sources are mostly self-evident from the captions. Finally, I would like to say thank you to my many friends and colleagues, in United Glass and elsewhere, who have provided information, advice and assistance in many other ways. I think it is true to say that the glass industry and the packaging industry have long realised the value of co-operation as a means of making progress, and of that fact this book is one small example.

B.E.M.

November 1977
United Glass Ltd, Porters Wood
St Albans

1. The place of glass in packaging

Glass is the senior member of the family of packaging materials, and has been so for over three hundred years.

In recent years it has been facing competition from several new materials, and facing it so well that the demand for glass containers in Britain has more than doubled since 1950. Fig. 1.1 shows the growth that has taken place in the eleven years 1966–76. Similar growth has occurred in most other European countries, and in the USA.

The number of glass containers produced for use in each country exhibits some interesting differences, as seen below.

Bottles per head of population in 1975

(Per capita consumption of manufacturers' sales)

USA	188
W. Germany	145
UK	116
Italy	105
Spain	69

One can draw a parallel between the number of bottles used and the standard of living in each country. It would be an unwarranted assumption to say that if the British standard of living reaches the present American level we shall necessarily be using the same number of bottles per head as they do, but at least the trend is likely to be in that direction. In fact the 27% overall growth in demand for bottles during the period 1966 to 1976 has outstripped the growth in real terms of both the Gross Domestic Product (20%) and consumer expenditure (22%).

Consumption figures in thousands of millions need a little breaking down before we can grasp what they really mean. The British consumption of over 6600 million bottles in 1976 is equivalent to more than 19 million per day. But many of these are multi-trip bottles, designed to be reused as many times as possible, so that the total number of bottles *filled* each day is much greater.

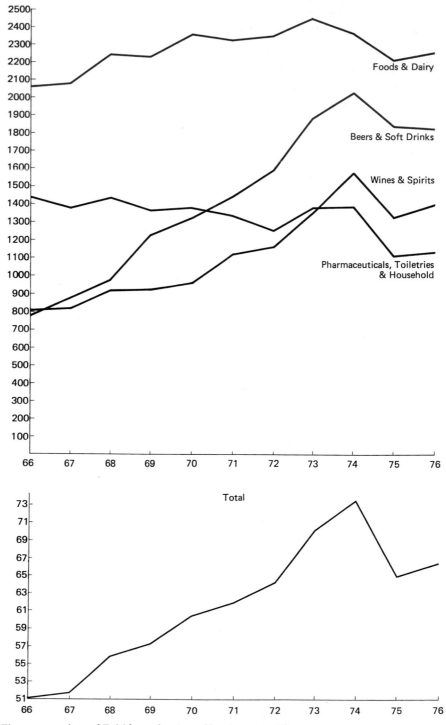

Figure 1.1.   Annual British production of bottles – in millions

Leaving aside fluctuations in stock levels, changes occur in the demand for bottles from any of three causes:

1. Increases or decreases in the sales of existing products.
2. Changes to glass from other types of container, and *vice versa.*
3. New products coming into use or old ones going out of production.

Examples of an increased use of glass in each of the three groups are:

1. Wines and carbonated minerals.
2. Baby foods.
3. Coffee 'creamers'.

As indicated in Fig. 1.1, the biggest growth has been in the large traditional markets for wines, spirits, beer and soft drinks. Use of glass for foods has remained fairly static, in line with food consumption itself (although changes in public taste and variations in disposable income can cause quite large increases or decreases in the consumption of particular kinds of foods).

The market areas where competition from alternative materials has had most impact are toiletries, cosmetics and household products; but the decline in the number of bottles sold to these industries has been more than counterbalanced by increased demand from other markets.

The overall UK growth pattern during the past ten years has followed the trend established during the immediate post-war period. During the whole of this time, however, some fluctuations in demand have occurred with the rise and fall in the economic fortunes of the nation as a whole. For example, during 1972 and 1973, real disposable income was growing very rapidly. Demand for packaged goods increased, and the result was that sales of glass containers jumped by over 10% in 1973. As the growth in disposable income fell away during 1975 consumer demand for all packaged goods began to turn down, and this partly explains the drop in demand for glass (and other packaging materials) in 1975. Another contributory factor was a build-up of stocks by packers in 1974 and subsequent destocking the following year. Demand for glass containers in 1976 turned up again, following the start of UK economic recovery in that year. Variations of this kind will inevitably occur again from time to time in the future, but the total demand for glass containers from 1977 onwards is expected to follow the steady growth pattern established prior to 1973.

Evidently, glass continues to be found suitable for a very wide range of applications, even though the packer is presented with an ever-increasing list of alternative materials. But many of these alternatives have been making good progress too, and clearly there is no sign yet of the elusive 'perfect' material which could be adopted universally.

So the packer still has to decide on his objectives and requirements for any new pack, and then work out which material will come nearest to meeting them. Let us consider what the general requirements for a successful pack are likely to be, and which characteristics of glass are most likely to be found relevant in such investigations.

From the packer's point of view, any container may need to satisfy a number of different criteria, most of which may be classified under three main headings: (1) technical (2) commercial, and (3) marketing. The technical needs are those which concern the efficiency of the pack in protecting and preserving its contents; the commercial criteria are those which relate to the overall cost of the pack; and the marketing needs are the ones concerned with the contribution made by the package to the promotion and sale of the product in a competitive retail market.

The relative importance of these three sets of criteria will naturally vary depending on the nature of the product. In earlier times, technical and cost factors were usually of paramount importance; but today the ability of the pack to 'sell itself' in the market-place may be of equal or greater significance – especially with the spread of supermarkets and self-service stores.

We will now take a closer look at all the relevant needs that can affect the use of glass from the packer's point of view, and follow by considering the consumer's needs, which it is just as important to satisfy if the initial success in selling is to be repeated indefinitely.

Technical needs

The following are the main characteristics of glass which are relevant to protecting the contents of the pack from the time it leaves the filling line until the bottle or jar is finally discarded by the consumer.

PRODUCT COMPATIBILITY

One of the strongest claims of glass containers is that they enable practically all liquids and solids (other than hydrofluoric acid) to be stored indefinitely without any effect on quality or flavour. The only proviso is that a suitable colour of glass (e.g. amber or green) should be used if the contents are light sensitive.

A detailed discussion of the chemical resistance of glass is given in Chapter 4, but for all practical purposes packers who use glass do not have to worry about whether their product will react chemically with the container, or whether undesirable or toxic chemicals will leach out from the glass into the contents. The same can not be said of many other packaging materials.

IMPERMEABILITY

In addition to being chemically inert, glass is absolutely impermeable to all gases and liquids. A scientist should always hesitate before using the words 'perfect' or 'absolute', but this is one case in which they are justifiable: no one has ever detected any diffusion of gas or liquid through the walls of a glass container in service. Diffusion or leakage is not possible unless the bottle is inadequately sealed. This characteristic can be of major importance in packaging highly volatile products containing alcohol or essential oils. It is also important in maintaining the quality of aqueous products on the one hand and hygroscopic products on the other.

ODOUR RESISTANCE

The problem of odours transmitted by packaging materials has come in for a good deal of investigation in recent years, especially in relation to food packaging. Here again, glass presents no difficulty. Not only is it completely odourless, but its impermeability ensures that odour is not transmitted to the contents from external sources. Even the most delicately flavoured foods are thus protected against any risk of tainting.

EFFECTIVE SEALING

Bottles or jars can be sealed in many ways to suit all kinds of products. Some closures provide a 'normal' seal (i.e. one where the pressure inside the bottle is roughly the same as that outside). Others provide a vacuum seal (designed to maintain a lower pressure inside the bottle) or a pressure seal (designed to withstand high internal pressures). Vacuum seals are used to preserve the quality of heat processed foods. Pressure seals are best known in the form of closures for carbonated beers and soft drinks. A full discussion of closures and sealing methods is given in Chapter 9.

B

STRENGTH

Glass is often thought of as being brittle and fragile. But there are many packaging plants where glass containers undergo filling and complex processing procedures at very high speeds and where breakage is virtually unknown. What are the real facts?

The first is that glass is potentially a remarkably strong material – perhaps twenty times stronger than steel. This theoretical strength is, however, not retained in practice because of the formation of sub-microscopic surface flaws in the glass soon after manufacture. In recent years, glass container manufacturers have devoted a lot of time and attention to discovering ways of utilising more of the theoretical strength of glass, and significant progress has been made. So much so that many glass containers today are twice as strong as their counterparts of a few years ago.

The mechanical properties of glass containers are fully discussed in Chapter 4. From the packer's point of view, there are three main hazards to consider. The first is the effect of impacts, and it cannot be disputed that glass is less resistant to impact than some of the other container materials. Usually risk of breakage can be reduced to negligible propor-tions by designing the container to give it as much strength as possible and by guarding it against uncontrolled impacts. This is quite easy for single-trip bottles, which spend the more critical part of their lives under the direct control of the packer or distributor. It may be more difficult to achieve with multi-trip bottles, but even so a great deal can be done by well-planned handling, and losses due to breakage can be kept down to very low levels.

The second hazard occurs when a liquid-filled container is dropped, causing a combination of internal pressure and external impact. In general a glass bottle can withstand this better than a treated paper container and about as well as a rigid plastic container, but not as well as some types of metal container.

The third circumstance to consider is that of internal pressure or internal vacuum: glass containers can withstand these forces as well as or better than any other types of container material.

DIMENSIONAL ACCURACY

Modern methods of glass container manufacture allow the production of bottles and jars within tight dimensional tolerances. Without such strict control over dimensional variations, filling and capping at today's high production speeds would be impossible. Details of recommended working tolerances are discussed in Chapter 3, while methods of checking tolerances are described in Chapter 6.

HYGIENE

Glass containers have three important characteristics which should be borne in mind whenever a really hygienic container is being sought, e.g. for foods or pharmaceutical products.

First, the degree of sterility of the inside of a new glass container is remarkably high – higher than can be achieved by any of the usual commercial processes of washing or 'sterilising' carried out on any other material. The reason is simply that when a glass bottle is made, the inside surface is cooled from the very high melting temperature without being touched by any solid or liquid. The only possible source of contamination is the small amount of air which enters the bottle neck as it cools, but fortunately the hot atmosphere surrounding the bottles during forming and annealing is inimical to bacteria.

Secondly, if the glass does become dirty or contaminated in use, the hardness and smoothness of its surface and its chemical inertness make cleaning processes effective and reliable. Dirt and bateria can survive cleaning if there arc pores or crevices in which they can hide, and if a contaminant reacts chemically with a container surface it will be very difficult to remove; neither of these hazards is likely with glass.

Thirdly, glass can easily withstand high-temperature sterilisation.

Commercial needs

The commercial and economic advantages of packaging in glass derive from five main characteristics: availability; low cost; returnability; ease of packing; and ease of distribution. These are discussed in detail below.

AVAILABILITY

There are at present about 30 glass container factories in England and Scotland, operated by 10 competing companies. The factories are to some extent grouped in the traditional glass-making areas, namely Lancashire (St Helens), Yorkshire, Central Scotland and London, and in each of these areas at least two competing manufacturers are represented. Some of the smaller companies do not make the full range of colours and make only a restricted range of designs; but no matter in which part of the United

Kingdom a packer may be located, there is always at least one glass container manufacturer who can conveniently supply whatever is needed.

Orders for very large numbers of bottles normally present no problem to an industry which produces over 19 million a day. The bottle maker does, however, need as much notice as possible of his customer's requirements so that he can plan his production accordingly.* Glass making is a continuous round-the-clock process and machine scheduling must be carefully arranged to achieve maximum efficiency in production.

If a new design of bottle is needed in quantity, a period of the order of three months may be needed by the manufacturer for his design work, trial production of glass samples, and for making a complete set of moulds. (For details of this work see Chapter 5.) Production of an established design of bottle, or one involving a small design change, can usually be arranged more quickly.

LOW COST

Visitors to a modern bottle factory, with all its complexities of furnaces, machines, annealing lehrs and engineering shops, often express surprise that bottles can be made so cheaply. (Bottle buyers usually manage to conceal this feeling.) The simple explanation, of course, lies in mass production, carried on continuously for 24 hours a day. The cost of a bottle depends on its glass content and complexity of shape, which control how fast it can be made, and on the number required, which determines the effect of the overhead costs.

It is not easy to maintain stable prices when fuel, labour and material costs are all rising. Nevertheless, British bottle makers have not been unsuccessful in this direction. Economies in the methods of making glass and improving methods of forming bottles have all played their part. The result is that bottles can be sold today for the same nominal price as they were 300 years ago. In those days a plain bottle cost 2 or 3 pence, but if embossed with the owner's name or crest the price was about 6 pence. In terms of real money these prices were, of course, much higher than modern ones, but it is an interesting coincidence that the decrease in bottle costs has more than kept pace with the decrease in the value of money.

---

* Many bottlers will remember that during 1973 and 1974 bottle shortages occurred because of a sudden upsurge in demand for all kinds of packaging materials as discussed earlier. Glass makers endeavour to forecast demand as accurately as possible but since new capacity can take three years to build, there is no way of producing more bottles overnight to cope with very large unexpected increases in usage.

In terms of cost-competitiveness with other rigid or semi-rigid packaging materials, glass has consistently improved its position in recent years. Although all the commonly used packaging materials have increased in price, the increase in glass container costs has been at a significantly lower level. One of the main reasons for this is that the raw materials of glass manufacture are all indigenous to Britain. They are plentiful, cheap and tend to be isolated from the political and other uncertainties which have affected materials derived from overseas sources. Added to this is the beneficial effect of the very heavy investment by glass container manufacturers in improved production facilities and technical expertise. Glass is therefore expected to maintain or even improve its competitive position in relation to other materials in the years ahead.

RETURNABILITY

The ability of glass bottles to be returned to the packer for reuse gives them an important economic advantage in a number of industries, of which the dairy, brewing and soft-drink trades are the most notable examples. The national *average* life of a milk bottle is about 25 trips from dairy to doorstep and back again, and some milk bottles make 40 trips or more. Beer and soft-drink bottles make an average of 15 trips, with beer bottles making rather more journeys than soft-drink bottles.

EASE OF FILLING

Filling bottles is basically a very simple process. It can be done by hand, by semi-automatic equipment, or by fully automatic machinery, as discussed in Chapter 8. Some modern filling lines now operate at up to 1000 containers per minute in the UK, while higher speeds are quite feasible.

In handling glass containers on these high-speed lines, the strength and rigidity of glass is a vital asset. It enables the containers to travel well on fast-moving conveyers and to be manipulated by complex filling and capping machinery.

EASE OF DISTRIBUTION

There is no difficulty in transporting glass containers by any of the usual methods, although in UK practice most bottles travel by road transport.

A lorry will usually be able to carry the same *volume* of containers regardless of whether they are light or heavy; compactness is then the main criterion of cost. Under these circumstances, short-necked bottles or wide-mouthed jars which waste very little space may be preferred. The need for compactness may sometimes be overruled by marketing considerations (such as the need for a distinctive package) and glass can provide any compromise which may be needed. Whatever the container shape, new handling techniques such as bulk palletising (see Chapter 7) permit much more compact and easily handled loads than in the past.

It might be thought that containers with cylindrical rather than square cross-sections would be wasteful of space. In fact, the loss is quite small: if cylindrical containers are placed in rows, for example, in a carton with square divided compartments, the amount of space wasted is 21%. If they are closely stacked as in a bulk pallet load or on a shelf, the lost space can be reduced to 9%, as shown in Fig. 1.2. This refers strictly to the *area*

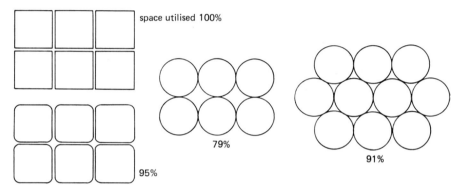

Figure 1.2.   Shelf utilisation with different container shapes

on which the containers stand; a calculation of volume utilisation would have to take account of the vertical shape.

The thickness of the glass wall has some slight effect on compactness, and in all cases where compactness is especially important this will be an additional reason for wanting to keep the wall as thin as practicable.

Marketing needs

Now let us examine the three main criteria which have to be considered in relation to containers which are displayed for sale to retail purchasers.

RECOGNITION

Recognisability of the product is the first of the marketing needs, and it is hardly necessary to emphasise the advantage of a transparent container in this respect. But the form of the container is important, too. The established glass designs for such products as jam, liquid medicines, tomato sauce, wines, beers and milk can all be recognised easily, even by a child (and in some cases also by domestic pets!)

Ideally the potential customer should be able to recognise instantly both the type of product and the brand, without having to read the label. Here, the variety in design which can be obtained with glass provides limitless opportunities for the creation of containers with individual brand identity. Fig. 1.3 shows a number of well-known bottles without their identifying labels: in one case the bottle is so distinctive that no label is normally used.

By careful modification of a traditional design it is often possible to obtain both product recognition and brand recognition together. Some of the latest designs of proprietary wine and spirit bottles are good examples of this. They are basically of traditional shape, but with suffi-

Figure 1.3. Distinctive bottle designs

cient difference to catch the eye. It is now becoming more and more common to see advertising campaigns for new products in which a distinctive pack is featured as part of the overall promotional scheme.

The possibilities of individuality in design do not, of course, preclude the fact that in some market areas there are even greater advantages to be obtained from standardisation, especially with returnable containers for products such as beers and milk. Standard or 'general trade' containers also play an important role in certain other markets, notably for wines and spirits, jams, pharmaceuticals and soft drinks, as discussed in Chapter 5.

CONTAINER APPEAL

Allied with product recognition and brand recognition is the fundamental appeal of the container to the consumer. The visual properties of glass – transparency, smooth-flowing shapes and brilliant surfaces – have always interested artists and designers, and even such a mundane subject as the design of a new bottle can excite a good deal of public interest. So it is no exaggeration to say that the use of well-designed glass containers can be of commercial advantage from the marketing point of view.

What about the image-creating properties of glass containers? Designers and psychologists have a great deal to say about this, involving the use of terms like 'physiognomic perception'. Discussion on these lines would perhaps be out of place in a simple technical account of bottles, but it is certainly an important subject, and a few examples of some of the more obvious stratagems may be mentioned:

1. The contents can be made to look cool and refreshing by giving the container surface a frosted appearance.
2. An impression of being tightly packed – presumably with something good – can be given by making the sides bulged or barrel-shaped.
3. The container can be made to look 'handy', and therefore useful, by giving it a shape which fits into the hand.
4. The contents can acquire an extra impression of value and high quality by giving the container flat facets such as are found on jewels and crystal glass.
5. The contents can be given an ultra-modern look, or alternatively, the appearance of being old-established, reliable and mature, by streamlining or by reproducing the shapes of the 'good old' bottles.

Whether or not any of these methods are employed, it is known from consumer surveys that the public generally associate glass with the idea of purity and cleanliness, and in their expectation of high quality they are likely to associate both the container and its contents. This brings us conveniently to the third marketing need.

PRODUCT APPEAL

Product appeal applies mainly to food and drink, but also extends to things like cosmetics. Obviously, if the product itself looks good the customer should be able to see it to best advantage and so a transparent container is desirable. In addition to quality, the customer is often encouraged by a visual indication of quantity, so again a transparent package helps. Some time ago an official campaign in the USA against 'deceptive packaging' was directed almost entirely against opaque packages which do not contain as much as they should. All this was summed up long ago in the proverb 'Don't buy a pig in a poke'.* (Chaucer's Pardoner in *The Canterbury Tales* knew the value of displaying what he was selling – 'and in a glas he haddé piggés bonés'.)

There is another important reason why food and drink may have additional appeal when packed in glass: this is concerned with the subtleties of flavour. Good flavour is not just solely a matter of preserving the product in good chemical condition: an element of psychology is also important. When the human brain is deciding on the flavour of something, it takes into account not only the information supplied by the tongue, nose and the other senses, but also any previous knowledge, opinion or association of ideas.

Preconceived ideas about the container can play an important part in some people's minds. The public generally expect products in glass to taste good, and they seem to remember with quite remarkable persistence the occasions when products packed in other ways were not good. This is undoubtedly one of the reasons why many of them believe that 'it tastes better from a bottle'.

Consumer needs

The main concern of the consumer is that the product should be preserved in good condition until it is required at the point of use. In this respect, consumers are just as interested in the various technical needs of a package enumerated above as is the packer, although in practice they will usually tend to take such matters very much for granted. What is of more immediate concern to them is the ease of using the product and the ease with which the pack may be resealed if the contents are not used all at

---

* A 'poke' is an opaque sack.

once. Both these needs are discussed below – followed by a note on after-use containers, which if not exactly a consumer 'need' are certainly a consumer benefit.

Quite apart from such essentially practical matters, consumers are also becoming increasingly concerned about the environmental issues associated with the use of different kinds of packaging materials, and these are discussed in detail in Chapter 14.

EASE OF UNPACKING AND RESEALING

The public have for many years gamely struggled with all sorts of tools to open their packages, but in the press-button age they are becoming much more critical and outspoken about difficult packs. It is hard to conceive anything simpler and easier than the screw closure (or 'twist-off' cap) for glass containers; several specialised types of this have been developed for all manner of liquids and solids (see Chapter 9), and they are becoming more and more popular.

A second advantage of screw closures is the ease with which they can be reapplied to give a good degree of protection to material remaining in the container. This feature is particularly appreciated by housewives for many products such as instant coffee and jam, which are used a little at a time.

For bottles which are emptied in one serving, there will probably always be a place for the simpler lever-off closures. The crown cork provides a remarkably simple, effective and easily removed closure for beer and minerals, and though it has been modified in minor ways over the years, it is likely to have a long life ahead of it yet.

Next, we should consider getting out the contents. For dry powders or solids which are removed with a spoon, the container needs a sufficiently wide neck and a smooth interior without ridges, ledges or overhangs to trap the contents. The outside, especially around the neck opening, should also be smooth and free from ledges or 'gutters'. Glass can satisfy these requirements very well.

The pouring of thin liquids presents few problems with any type of container; glass usually offers an extra advantage in that the surface over which the liquid flows is normally protected from dirt while the closure is in position.

Pouring of thick liquids such as oils, syrups and sauces is more difficult to achieve cleanly, and some glass containers are by no means perfect in this respect. It is not difficult to make a bottle finish with excellent pouring characteristics, but if the 'ideal' pouring lip has to be modified to give greater mechanical strength or better sealing ability, then the pouring characteristics may be compromised.

AFTER-USE CONTAINERS

Bottles and jars have for centuries been used by consumers for storing all kinds of articles long after the original contents of the container have been used up. Today, this secondary use of glass containers finds its highest expression in the special after-use packs which have been employed so successfully for packaging a great many different kinds of foods and household products.

Perhaps the best-known examples of after-use packs in the United Kingdom are the storage jars now widely used for packaging instant coffee. Other attractive storage jars have also been used for packaging products as varied as boiled sweets and bath salts, while glass tumblers have been used for packaging coffee and jam.

In other parts of Europe after-use packs are even more widespread than in Britain. In fact, they have even become the 'normal' pack for many products. Drinking glasses and beer mugs are especially popular, and the products packed in them range from mustard and tomato purée to cocktail cherries and stuffed olives.

2. How bottles are made

Glass containers are so much a part of everyday life that most people take them completely for granted, never stopping to think how they are made. Some of those who do wonder about it produce very curious ideas – one suggestion not uncommon is that they must be made in two halves which are then stuck together. (This is how glass blocks are made, but not bottles.)

It is unlikely that glass packers suffer from such misconceptions. Several good descriptions of container manufacture have been written for their benefit, and there are also films on the subject made by the Glass Manufacturers' Federation and others. So the aim of this chapter is not to describe the whole of bottle making once again, but to pick out those points which may be most helpful or interesting to the glass packer. An understanding of some of the problems and techniques of glass making should help the packer to appreciate better the capabilities and limitations of glass, and so to make the best use of it.

Raw materials and glass compositions

There is a traditional account of glass composition which says that bottle glass would have been made of pure silica if its melting point had not been so high (1710°C), and that soda had to be added to 'lower the melting point'. Then, because this glass was soluble in water, lime had to be added to 'stabilise' it.

This is simple and sounds scientific but it is not quite honest. The first glass was made nearly 10,000 years ago, and was used for coating beads, long before there was the slightest understanding of chemistry or temperature measurement. Glass must have been discovered by accident – Pliny tells a story of the Phoenician sailors who used blocks of natron (natural soda) to support their cooking pots over fires on a sandy beach, and this is as good a suggestion as any. There were no doubt many different glass recipes in Roman times, including such ingredients as wood ash and burnt seaweed. Some of the glasses were probably not very durable and have long since disappeared, but there are many examples which have survived two thousand years without deterioration. The London Museum shows some fine examples of the clear glass jars used by the Romans to preserve their cremation ashes.

Only during the last sixty years has the science of glass become established, and now that the effects of all the possible ingredients are known, it is possible to design a glass composition 'from first principles'. Nevertheless the compositions used for modern glass containers bear a surprising similarity to some of the best Roman ones! The reader may be tempted to draw various conclusions from this surprising fact; the important point, of course, is that modern container glass can be relied on to last for at least 2000 years.

Glass is a mixture in which all the constituent atoms have been persuaded by heating to link up into a random but rigid network, in which each silicon atom is linked to four oxygen atoms and thence to other silicon atoms, with atoms of sodium and calcium distributed in the holes in the network. It is customary to express the composition as if the glass were a mixture of oxides. The example shown below gives a typical modern British glass composition, together with the common names used for the oxides, and notes on the raw materials used.

COMPOSITION (BY WEIGHT) OF A MODERN CONTAINER GLASS:

Silica	(SiO_2)	72%	from high purity sand or quartz rock.
Lime	(CaO)	11%	from limestone (calcium carbonate).
Soda	(Na_2O)	14%	from soda-ash (synthetic sodium carbonate).
Alumina	(Al_2O_3)	1·7%	Some alumina is present in the other raw materials. The rest is usually obtained from a feldspar-type aluminous mineral, or from processed blast furnace slag.
Magnesia	(MgO)	0·3%	not usually added intentionally, but are acceptable impurities in the other materials.
Potash	(K_2O)	0·3%	

This composition has been established as the best and most economical possible from the raw materials available in this country, and the glass has properties equal to or better than any produced elsewhere. In other countries some compositions are slightly different, because of different raw materials. Where the available limestone contains magnesia as well as lime (dolomite) the glass may contain about 3·5% of magnesia and only 8% of lime. In some parts of North America the silica sands contain several per cent of alumina, and when they are used the silica and alumina contents of the glass will change accordingly.

Glass colours

Most British containers are made of colourless glass, but there is also a range of colours available, amber being the most widely used.

Colourless or 'white flint'.   If all the major raw materials were perfectly pure the glass produced would be completely colourless. In practice some impurities are unavoidable, and the content of iron oxide (Fe_2O_3) cannot usually be lowered to less than about 0·04%. This would give a definite blue-green tinge to the glass, but ways have been found of neutralising this slight colour by adding small amounts of 'decolorisers', usually containing selenium. Decolorisers will not work if the iron oxide content is over about 0·08%.

Pale green.   If slightly less pure materials are used the iron oxide content may rise to about 0·15%, and pale green glass is obtained. This can be a blue-green colour, similar to that of thick window glass, but if the iron is oxidised, by adding oxidising agents to the raw materials, the colour becomes yellower.

The blue-green colour can be slightly deepened by adding a trace of chromium oxide (green), to produce what the Americans call 'Georgia green'.

Green.   This colour, too, is obtained basically with a combination of chromium oxide and iron oxide.

Amber.   The familiar brown colour, commonly used for beer bottles, is called amber. It is obtained by melting a glass of moderate iron oxide content under strongly reducing conditions. Usually some carbon is added to ensure this, and there must also be a little sulphur in the glass.

Blue.   If about 0·1% of cobalt oxide is added to a low-iron glass, an intense blue colour is obtained.

Glass melting

To produce glass in large quantities, as required for automatic bottle machines, a furnace shaped roughly like a small swimming bath is used, with a surface area of up to about 100 m² and a depth of 1·5 to 2 m of molten glass. The finished glass is withdrawn continuously from one end, while raw materials are fed in at the same rate at the other end. Most furnaces are now heated by oil or natural gas which is injected into the space between the glass and the furnace roof. This may seem a curious arrangement to the layman who thinks in terms of standing a kettle on top of a gas ring, but at the temperature of 1500°C or more required for glass melting no other way is feasible. Electric heating is possible technically, as molten glass conducts electricity quite well, but as a main source of heat it would (at present prices) considerably increase the cost of glass making. A little electricity is used to 'boost' some furnaces, and used in this way it can be an economic proposition.

The secret of good glass melting is to keep all the conditions as steady as possible. Once the furnace is up to temperature it is kept running continuously until the end of the campaign, and even when no glass is being withdrawn the temperature must be maintained. The word 'campaign' is very aptly used to describe the life of a furnace, as it really does become a battle to keep the furnace in continuous operation for as long as possible. The glass 'bath' is lined with thick walls of refractories specially developed to withstand the intense corrosive action of molten glass, and these make it possible – at a price of up to about £3500 per m³ for the refractories – to run a furnace continuously for over seven years. During this period the output from a furnace of average size, say 65 m² in area, might approach half a million tons of glass.

The principles of bottle making

In any discussion of bottle making it is impossible to avoid using some of the bottle maker's jargon, so we will start by explaining what the essential terms mean. In addition the names given to the main parts of the bottle are shown in Fig. 2.1.

Flow machine: A bottle-forming machine to which molten glass is supplied as separate gobs, which fall from an overhead feeding device. The gobs are roughly sausage-shaped.

Blank mould, sometimes called parison mould: The mould in which the first stage of shaping the body of the bottle is carried out.

Neck ring: A separate mould which forms the *finish* of the bottle, i.e. the top part of the neck including all the details of form required by the closure.

Parison, sometimes called blank: The partly formed bottle. At this stage the neck and finish are in their final form, but the body is not yet shaped.

Blow mould: The mould in which the formation of the bottle body is completed.

There are four essential stages in automatic bottle making:

Gathering: The required quantity of glass is obtained from the furnace.

Forming the parison: This shaping process may be carried out by pressure, and/or suction, or by a metal plunger.

Transfer: The parison, held by its neck, is transferred from the blank mould to the blow mould. All the moulds are made in two halves, which open on hinges to allow the transfer to occur.

Blowing: Compressed air applied inside the parison forces the glass out to the shape of the blow mould.

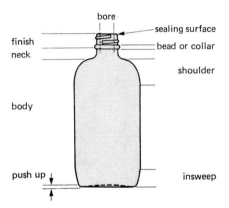

Figure 2.1. Naming the parts

This sequence is the same as in the old hand-methods of bottle making, except that the finishing of the neck used to be the final operation, after the bottle had been detached from the blowing tube. This is why the bottle mouth is still called the 'finish'.

The sequence of operations is illustrated diagrammatically in Fig. 2.2 for two different types of forming operation. The essential difference between them is in the method of forming the parison, but there are other differences and it is likely that for any particular type of bottle one of the methods will be better or cheaper than the other.

Let us first consider the gathering of the glass, which is basically the same for both processes. In order to get the internal capacity of the bottle right it is necessary to put the correct volume of glass into a mould cavity of the correct volume. Glass from the furnace flows downwards through an orifice in a long trough, assisted by a reciprocating plunger above the hole. A pair of shears is operated at regular intervals to chop the glass into a series of individual 'gobs', which then fall into the blank moulds, either vertically or via an inclined chute. The amount of glass obtained depends on the temperature and composition of the glass (both of which affect the rate of flow), and on the size of the orifice, the length of stroke of the plunger, and the timing of the shears.

Now let us consider the shaping of the parison. In the 'blow-and-blow' process (Fig. 2.2), the gob is dropped into the blank mould, then two puffs of compressed air are successively applied at each end of the parison. This process allows a wide choice of parison shape and a correspondingly wide choice of final container shape. At transfer the parison is inverted from the opened blank mould into the opened blow mould. The latter closes and the neck ring opens, leaving the parison hanging from the top of the blow mould, supported by its finish bead.

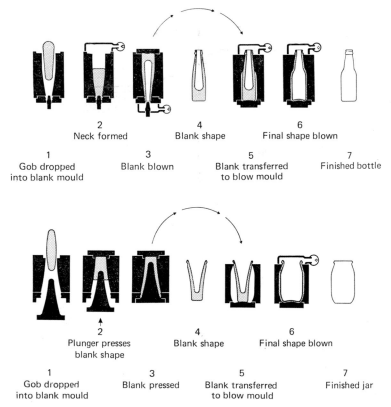

Figure 2.2. Bottle forming: (Above) Blow-and-blow; (Below) Press-and-blow

C

In the 'press-and-blow' process the parison itself is made not by blowing but by being pressed into an exact shape by a plunger; this makes the process especially suitable for wide-mouthed jars.

The final operation – blowing the parison into the shape of the blow mould – is essentially the same for both processes.

There are, of course, some variations on this theme with certain types of machine. For example, in the Roirant machines (R.7 and S.10) the final formation of the bottle is achieved by sucking the glass out to the mould instead of by blowing from the inside. The final result, however, is basically the same.

A complete bottle-forming machine consists of several separate stations, each with its neck ring, blank mould and blow mould. The way in which the stations are arranged varies considerably. In some machines all the blank moulds are on one rotating table, and all the blow moulds on another; others have both sets of moulds on a single rotating table. Another type, the *individual section* or IS machine, has no rotating table, but the sections, each consisting of a blank mould and a blow mould, are arranged in a row, and operate independently of each other. This is now the most widely used machine in the glass container industry.

An eight-section Emhart IS machine is shown in action in Fig. 2.3. It will be seen that the finished containers are emerging in pairs. In this 'dual' operation each station has twin neck rings, blank moulds and blow moulds, and the station is fed with twin gobs falling side by side. With this system the output of the machine can be nearly doubled, so obviously it is used wherever the dimensions and design of the bottle, and the amount of glass available from the furnace, permit. The limiting factor is usually the bottle diameter, as on any machine the distance between the centres of the twin moulds is fixed.

For some smaller and simple designs of bottle 'triple-gob' operation is likely to become more feasible in future, and even quadruple operation has been achieved for some tiny bottles.

Bottle making in practice

So much for the theory of bottle making; what does the bottle maker have to do to ensure that he makes the best possible bottles?

Assuming that the basic bottle design is known, the first thing to do is to choose the right machine for the job. The choice will depend on the characteristics of the bottle, and also the numbers required. Large machines are most economic for large-scale production.

The next step is to design the parison; this is the starting point in getting

Figure 2.3. Eight-section IS machine

the best glass distribution. If the bottle design is very irregular or complex it may be impossible to design a parison which will give the desired glass distribution, and the only solution may be to make the bottle unnecessarily thick in some places, to ensure that it is not unacceptably thin in others.

The blank and blow moulds now have to be accurately machined, usually from fine-grained cast iron. The dimensions and volume of the blow-mould cavity have to be checked, and the machined surfaces polished to nearly a mirror finish. (The accuracy attained in mould making, and its effect on bottle dimensions and capacity, will be discussed in Chapter 3.) Any embossed lettering or decoration required on the bottle is obtained by engraving the mould surface. There are, of course, other industries for which accurate moulds have to be made, but only in bottle making is the mould required to produce an article to exact dimensions when the mould is heated to around 500°C, and the initial temperature of the moulding material is about 1000°C. (A complete set of moulds, and equipment for automatic mould making, are shown in Figs. 2.4 and 2.5.)

If the parison design is satisfactory, the moulds accurate, and the right amount of glass is being delivered at the right temperature, then it is possible to make a good bottle. The machine has two basic jobs to do: the first is to shape the hot glass, and the second, equally important, is to extract the heat as quickly and uniformly as possible. Forced air cooling, and sometimes water cooling, is applied to many parts of the machine and to the glass, and considerable skill is needed to do this to the best effect. If the moulds become too hot, the glass may stick to the mould and tear when pulled away; if the glass or the moulds become too cold,

Figure 2.4. Complete set of mould equipment

the glass will not flow properly into the final position. An unnecessarily complex design of bottle neck and finish can be a great hindrance in this respect; the designer should always bear in mind that glass enters the mould from the bottom and has to be squeezed up into the far end of the neck mould.

When the bottle leaves the forming machine its outer surface is rigid,

Figure 2.5. Automatic mould cavity machining at Johnson Radley, Pudsey

having cooled to perhaps 300°C, but the inner parts are still hot and soft. If it were to continue cooling naturally, the hot inner parts would contract more than the cool outer surface, and dangerous stresses would be set up, the inner parts being stretched and the outer surface compressed. To avoid this the bottles have to be annealed; they are reheated to about 550°C, at which temperature the glass can flow very slightly and relieve any stresses, then cooled sufficiently slowly to prevent further stresses being set up. (The long tunnel in which this reheating and slow cooling occurs is called the *annealing lehr*.) A small amount of compression in the outer surface is an advantage because it has a slight toughening effect, provided it can be achieved without danger of a corresponding weakening effect at the neck or inside surface. So in some cases the bottle maker adjusts the annealing rate to obtain a controlled amount of compression. The expression 'dis-annealing' has been used to describe this process.

CONTROL OF DIMENSIONS AND CAPACITY

The bottle dimensions and the internal capacity are two properties of great importance to the user, and they are continually checked during manufacture. Extensive use is made of gauges for checking, and the fine detail of bottle finishes is checked on optical profile projectors. Also, as discussed in Chapter 3, dimensions are increasingly checked by automatic means. The bottle weights are checked continuously as they come off the forming machine, and this gives a good guide to what is happening to the capacities. Final checks are made after annealing; the methods used are described in Chapter 3.

BOTTLE INSPECTION

Every bottle made is carefully inspected after annealing, and defective ones returned to the furnace for remelting. Even when all operating conditions are as good as possible, there are always a few bottles to reject, because they are malformed or damaged or contain a refractory inclusion in the glass. As the bottle design becomes more complex, the proportion of defectives tends to rise, because the operating conditions, especially the temperatures, become more critical. Information about the nature of the defects is continually fed back to the operators of the forming machines who make any adjustments possible to cure the faults, and to maintain the highest production efficiency.

As with dimensions, inspection for defects is increasingly being carried out by automatic means, because it is becoming more reliable and

sophisticated than human inspection, and also because it is at least potentially less expensive. One should nevertheless not make the mistake of
thinking that mechanical inspection is ever 100% efficient: machines, like
men, sometimes have their personal idiosyncracies, and they all need
regular care and supervision – by men – to achieve the best performance.

Ancillary processes

SURFACE TREATMENT

A bottle may be given an internal surface treatment to improve its
chemical properties or an external treatment to modify its mechanical
properties. Internal treatments during bottle production are carried out
either by placing a tablet containing sulphur or ammonium sulphate
inside the bottle just before it enters the annealing lehr, or by direct
injection of acid gases or organic fluorides into the bottle at the same stage.
The effect of this treatment is discussed in Chapter 4.

External treatments are widely used to provide additional strength and
abrasion resistance, and generally comprise two treatments which may be
used either singly or in combination.

The first of these treatments is generally referred to as the 'hot end'
treatment because it is carried out soon after the bottles are made, while
they are travelling between the forming machine and the annealing lehr.
It generally consists in the application of a compound of a metal, either in
sprayed liquid or vapour form, causing decomposition to take place at the
surface of the glass, giving essentially a film of metal oxide. Tin compounds
are commonly used for this purpose, and may be applied either as stannic
chloride vapour, or – as in the Titanising process – as an organic compound
of tin in liquid form.

Certain other metal compounds can also be used, such as those of titanium and zirconium. In all cases the coating produced in this way is essentially part of the glass and cannot be removed by normal washing.
although repeated washing in an alkaline solution will gradually dissolve
the coating.

The second treatment is generally referred to as the 'cold end' treatment
because it is carried out when the bottles are close to or emerging from the
cold end of the annealing lehr. Nevertheless, even at this point the bottles
can be at a temperature of 100° C or more and in that sense the term 'cold
end' treatment may be a little misleading.

A great variety of cold end treatment materials are available, but
generally they fall into two classes : those soluble in water and those

insoluble in water. Examples of soluble cold end coatings include poly-ethylene glycols and their esters. These are generally applied in dilute aqueous solution, while insoluble coatings are applied as suspensions in water. In both cases application is generally made using a sprayhead that continuously traverses backward and forward across the annealing lehr. A more recent cold end treatment material is oleic acid, which is applied as vaporised droplets in a special treatment hood. The effects of these various treatments are discussed further in Chapter 4.

Bottles given an adequate standard of hot and cold end treatment are extremely resistant to abrasion by metal and glass. However, most insoluble treatments are not yet resistant to repeated washing with alkaline detergents, and surface treatments are therefore used predominantly on single-trip bottles.

ENAMELLING

The application of permanent vitreous enamel labels to bottles is not an integral part of the production process, but is carried out by the bottle maker as a separate operation. The coloured enamels are themselves glasses in powdered form, mixed with a binder to make them into a thick 'paint'. Each colour is applied separately to the bottle by means of a silk-screen stencil, and then the bottles go through another lehr, in which the temperature is raised to near 600°C, to fuse the enamel permanently to the glass. The cooling rate must again be controlled carefully, otherwise the annealing of the bottles might be upset.

The start of bottle production

We have seen that furnaces and bottle machines are capable of running continuously for very long periods. But when a machine has to change to a different pattern of bottle, production must be interrupted for about an hour while the moulds and other machine parts are changed. The moulds are pre-heated in ovens to get them near to their operating conditions, but it takes an hour or more of production time to achieve real stability of temperature. Many adjustments are needed, to the timing, the cooling, machine speed and so on, during the settling-down period; for a simple bottle this period may be only an hour or two, but for a difficult bottle it could be a day or longer.

The economic effect of this period may best be illustrated by a hypo-thetical example. Consider a machine theoretically capable of making bottles to the value of £200 per hour. In practice its production efficiency

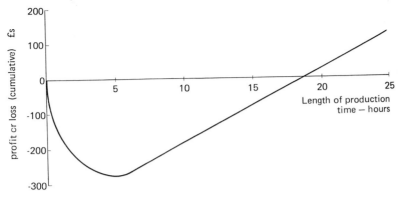

Figure 2.6. Profitability of a bottle-forming machine

would not reach 100% as some bottles are rejected, say, 10% of the total, so its actual maximum output would be worth £180 per hour.

The direct cost of operating such a machine and supplying it with glass might be £160 per hour, dropping to £140 when the machine is temporarily idle and not using glass. For a bottle of moderate difficulty the production efficiency might start at 50%, and might rise by about 10% an hour, levelling out at 90%. The effect of all this on the profitability of the machine is shown in Fig. 2.6.

It will be seen in this example that although the machine reaches its maximum efficiency in five hours it takes about 20 hours to wipe off the loss incurred in the early period. For a difficult bottle the unprofitable period could be much longer.

The costs of bottle making

We have seen above that it is not an economic proposition to run a bottle machine for short periods, and generally a run of about three days would be regarded as an absolute minimum. The output from modern machines may range from about 15,000 to 150,000 bottles per day, depending on the weight of the bottle and size of machine. So the minimum number of bottles which could be made in a single run could sometimes be as high as a million. This figure should be regarded as an order of magnitude only, as it depends on many factors, and a manufacturer may agree to make bottles in uneconomic quantities as a service to a customer who needs them urgently, perhaps for a market test on a new product. Alternatively the maker may produce more than is immediately required and put the rest in stock – this too brings its own economic problems.

As raw material costs are very small, the cost of a bottle largely depends on the speed and efficiency with which it can be produced. The speed is governed mainly by the time required to cool the glass to a rigid condition; this depends partly on the weight of glass and also on the shape of the bottle and the glass distribution. In general a light bottle can be made faster than a heavy one, but there are many exceptions, especially if a bottle with a light body has a heavy neck. Bottles of complex shape set special problems, and often the bottle maker has to deal with them by reducing speeds. Production efficiency may also suffer.

Another important item affecting the economics of bottle making is the cost of making, maintaining and repairing the moulds. A 'single' mould, i.e. the equipment required for one head on a machine, consists of at least nine separate parts (as shown in Fig. 2.4), and a complete set for an eight-head dual machine could cost between £5000 and £19,000. It is clearly vital that the bottle maker should be able to obtain a long working life from the moulds; a single mould may be capable of producing something like a million bottles before it has to be scrapped. Frequent design changes, and requests for unnecessarily tight bottle tolerances, put moulds out of service prematurely, and push up production costs.

We may therefore summarise the four basic requirements to be satisfied if a glass container is to be produced as cheaply as possible:

1. Demand for the bottle should be such that it can be made in large quantities.
2. The glass weight should not be greater than needed for strength and other requirements.
3. The design should be simple and symmetrical.
4. Dimensional tolerances should not be tighter than necessary.

It should be emphasised that here we have been discussing solely the costs of bottle making. When we come to discuss the technical and commercial considerations, including such matters as sales appeal, it may often be found worth while to look beyond the simple economic factors. Whatever design is chosen, however, someone has to 'count the cost', and the above section is intended to assist this vital operation.

3. Dimensions, tolerances and quality control

One of the assets of glass is the almost infinite variety of container design that it makes possible. There are about two thousand different designs in successful use today, although as we have seen in Chapter 2 the complexity of some of the designs has an effect on their cost. Some designs, too, may be doubtful propositions because they would be unsuitable for high-speed handling in the packer's factory. Apart from these considerations, it is unlikely that any suggested shape would be pronounced as completely impossible by a modern manufacturer, especially if he has a range of different types of forming machine available, and a wide experience of making all the possible shapes.

With regard to bottle sizes, any height up to about 380 mm and any width up to about 200 mm is possible with modern automatic forming machines. (It should be noted that in the United Kingdom bottle dimensions are invariably expressed in millimetres [mm] as recommended in British Standard 4602, *The use of metric units in specifications for glass containers and finishes.*) If larger ones should be required for special purposes, for example display or advertisement, it might be possible to make them by semi-automatic or manual methods.

Dimensional accuracy

Before discussing how much variation in dimensions is permissible, it is important to understand how and why the variations occur. The dimensional accuracy depends on:

1. How accurately the mould cavities are made.

2. How accurately the mould dimensions are controlled during the life of the mould.

3. How accurately the glass can be blown to fit the mould.

4. How accurately the bottle retains the mould shape while it is cooling immediately after leaving the mould.

We saw in Chapter 2 how the moulds are made from solid cast iron or similar metal, and 16 or more identical mould cavities have to be made to form a complete set for a large machine. The craftsmanship of the mould maker and the precision of machines like the one shown in Fig. 2.5 mean that even the most complex shape of mould can be made accurate to about 0·05 mm when new, and the parts of the mould which form the details of the bottle finish more accurately even than this. Not surprisingly, a simple cylindrical cavity can be made more accurately and more cheaply than any other shape.

While the mould is in action, its surface temperature is constantly varying between about 450 and 600°C, and the inside surface very slowly wears away through oxidation. Also it is not impossible for the whole body of the mould to warp slightly under these conditions. So the mould has to be taken out of production at intervals, to be cleaned free of scale or other deposit, checked for dimensions, and re-polished. There are several ingenious engineering tricks, such as spraying molten metal on to parts of the mould which have worn unduly, and inserting patches or 'bushings' at points of severe wear, but there is a limit to what can be done, and inevitably the dimensions of the mould cavity increase until, eventually, it has to be scrapped.

By comparison with the complexities of mould making, blowing the molten glass to fit the mould sounds easy. But there are snags. If the glass is just a little too hot, the bottle may not be sufficiently rigid when it leaves the mould, and if it is just a little too cold, it may set rigid before the last parts of the glass have been blown into place. The latter trouble is unlikely with bottles of simple symmetrical shape, because all the parts of the glass arrive at the mould surface at practically the same time. While the molten glass expands to shape, it has to displace the air which was originally in the mould. Fine escape holes and channels can be provided in the mould surface, but even so it is possible for small pockets of air to be trapped in the corners of irregular mould shapes. If all these dangers have been avoided, some parts of the bottle, especially any large flat panels, have a habit of warping inwards slightly during the cooling and annealing.

The four factors discussed above together cause slight dimensional variations from bottle to bottle, and in order to control these effectively, we first need to know the usual pattern of variation. Measurements on large numbers of any mass-produced article always show the same type of distribution, following the well-known bell-shaped or Gaussian curve. (For those who are not familiar with this, the curve is illustrated and explained in Appendix D.)

FIXING DIMENSIONAL LIMITS

It is all very easy for a supplier to say, 'We usually work to a limit of plus or minus so much', but this approach by itself does not give the purchaser much information about the range of variation likely in the goods he will receive. At the other end of the scale, a purchaser may say to the supplier, 'No single article you supply must ever exceed a length (or other dimension) of so much', but no manufacturer can hope to give an absolute guarantee of this sort, unless he spends a long time measuring every single article made. Fortunately neither of these situations is likely to occur today. It is generally recognised that modern automatic production methods and modern quality control can provide something much more precise than just 'working tolerances', and consumers recognise that it is not usually in their own economic interests to try to get a dimensional guarantee with every individual article from the production line.

It has been found that the most useful way of stating glass container tolerances is to use what are known as '2-sigma limits', indicating that theoretically at least 95·4% of all the containers produced should be within the upper and lower limits.

The statistical theory from which the 2-sigma limits are derived is discussed in Appendix D. Briefly, it asserts that if various statistical assumptions are satisfied, then 2·3% of all mass-produced articles will be above the upper limit and 2·3% below the lower one.

This approach does not, of course, mean that 23 bottles out of 1000 are likely to jam on the filling line because they are 'over-size'. The sensible packer, in collaboration with his bottle supplier, will work from the agreed 2-sigma limits to ensure that there is an adequate margin of safety on the filling line. Suppose, for example, that a packer is using bottles of nominal diameter 75 mm with a 2-sigma limit of ± 1·4 mm. He will, if wise, arrange for the clearances on the line to permit about double this amount of variation. Then, according to statistical theory, only one bottle in 30,000 should have a risk of jamming, and in real life the actual proportion would be even less than this.

There may still be some people who prefer not to discuss this subject, in the hope that they can preserve the illusion that it is possible to achieve perfection in a mass-produced article. It is, however, the declared aim of this book to give packers accurate and realistic information about bottles. So it should be understood that throughout this book all limits and tolerances refer to 2-sigma limits, unless stated otherwise.*

* Sometimes 3-sigma limits are used when an attempt is being made to make a count of the total number of notionally 'defective' bottles. This applies especially where verticality defects are grouped with other defects for quality control purposes, as described later.

The reader may wonder what happens if every bottle produced is put through an automatic gauging machine; cannot the manufacturer then achieve certainty? The answer is that it might well be possible to reduce the tolerances – at a cost, namely the cost of the gauging machinery and the cost of the bottles being rejected – but there would still be an element of variability in the operation of the gauging machine itself. So it would still be necessary to preserve the probability concept in the tolerance, but the statistics would become more difficult because the distribution of the surviving bottle diameters would no longer be Gaussian. Fortunately this approach has never been necessary in practice because it is quite easy to plan and operate fast and efficient filling lines so that they will accommodate the usual 'natural' range of dimensional variation.

DIMENSIONAL TOLERANCES IN PRACTICE

Filling speeds in the UK have increased steadily during the last 25 years, as shown by the following examples:

Type of bottle	*Filling speed achieved (bottles/minute)*	
	1952	*1977*
Small mineral	250	800
Half-pint beer	150	800
Pint milk	100	600

This advance could have resulted in unrealistic demands for bottles with tighter and tighter tolerances, but fortunately for all concerned this has not happened to an unreasonable extent. There has generally been good collaboration between makers and users, and some excellent filling and handling machinery has become available. Indeed, it would not be an exaggeration to say that the modern very high-speed filling lines are among the most efficient and trouble-free of any in the country.

During the last 10 years, British glass manufacturers have been working together, and with their customers, to establish realistic tolerances which will achieve adequate performance at realistic cost. Tables were first published by the Glass Manufacturers' Federation in 1965, revised in 1970 and 1974, for height, body, and verticality tolerances. Although they are in no way mandatory, they represent a realistic range of tolerances aimed at minimising the overall cost of making, filling and marketing glass containers. In other words, all glass container manufacturers should be able to produce containers to the specified 2-sigma tolerances at a realistic cost, and all the containers thus supplied should perform satisfactorily on high-speed packaging machinery.

The recommended tolerances for container height and leading horizontal body dimension are as follows:

TABLE 1: *Overall Height Tolerances*   TABLE 2: *Body Tolerances*

H (mm) Up to and including	T_H (mm) ±	D (mm) Up to and including	T_D (mm) ±
25	0·7	25·0	0·8
50	0·8	37·5	0·9
75	0·9	50·0	1·1
100	1·0	62·5	1·2
125	1·1	75·0	1·4
150	1·2	87·5	1·5
175	1·3	100·0	1·7
200	1·4	112·5	1·8
225	1·5	125·0	2·0
250	1·6	137·5	2·1
275	1·7	150·0	2·3
300	1·8		

Graphs of the nominal height and body dimensions against the tolerances listed here are essentially straight lines represented by the equations:

T_H = tolerance on overall height H
$$= \pm (0·6 + 0·004\ H)$$

and

T_D = tolerance on leading body dimension D
$$= \pm (0·5 + 0·012\ D)$$

For bottles of square, rectangular or oval cross-section the same tolerance is needed for all the relevant horizontal dimensions of the body. Thus, in the case of square or rectangular bottles, the leading dimension is the diagonal between opposite corners and the corresponding tolerance will apply also to the dimension measured parallel to either of the sides of the rectangle. For oval bottles, the tolerance for the minor axis will be the same as that for the major axis.

These figures should be regarded as a guide only; the tolerance required in a particular situation might occasionally be affected by the type of closure, the type of carton or crate used, etc. For bottles which are handled and filled at more moderate speeds, the above tolerances might be needlessly severe, but in every case the manufacturer will do his best to supply the packer with what he needs.

If the bottle maker and the packer were working for the same group, it would not be difficult to evaluate the level of bottle tolerances which would give the greatest overall profit, balancing cost of mould manufacture and glass production against any economies on the bottling line. The optimum tolerances for highest total profit are still the same if two separate firms are involved, and both packer and bottle maker should benefit by working together to put the most profitable pack on to the market.

VERTICALITY

When a bottle first leaves its mould it should stand perfectly upright, and if it is well annealed and cooled it should stay this way. It can happen, however, that during annealing the neck bends slightly or the base warps, throwing the bottle 'out of true'. If the defect is obviously visible to the eye the bottle will be rejected by the sorters, or by an automatic gauging machine, but if it is very slight and only detectable by measurement the sensible bottle maker may be uncertain whether the bottle is usable. In order to investigate this we must first obtain a measure of the deviation; the essential quantity is the horizontal distance by which the bottle finish departs from its intended position in relation to the base of the bottle. This distance is usually referred to as the 'verticality' of the bottle, but it would be more logical to call it 'non-verticality' or 'obliquity'.

With a cylindrical bottle the centre of the finish is always intended to be vertically above the centre of the base, and a measurement can easily be made by locating the base of the bottle in some form of self-centring chuck, and rotating it. The centre of the finish will move in a circle, of radius equal to the non-verticality. With a less symmetrical design of bottle, some form of template may be needed to locate the bottom of the bottle accurately. (It may be helpful to remember that the centre of the finish is always intended to be in the vertical plane through the joining faces of the body mould, the position of which is marked by the vertical seams on the side of the bottle.)

Lack of verticality can cause difficulty on filling lines if, as a result, the bottle neck is not correctly positioned beneath the filling head, or if the finish is not in the correct position or at the correct angle when capped.

The recommended 2-sigma tolerances for verticality have a constant value for bottle heights below 120 mm. Thereafter, they increase linearly according to the equation:

$$T = \text{verticality for overall height } H$$
$$= 0\cdot3 + 0\cdot01\ H$$

The resultant tolerances are as follows:

TABLE 3

Height H (mm) Up to and including	Verticality T_v (mm)
120	1·5
150	1·8
175	2·1
200	2·3
225	2·6
250	2·8
275	3·1
300	3·3

The figures quoted do not mean that, say, a 150 mm bottle which is just on the 2-sigma limit will be presented to the filling head with its neck 1·8 mm out of line. Bottles are rarely positioned solely by means of their bases, as the guides, starwheels or other system used to position the bottle generally contact a bottle somewhere above its mid-point. So this 'extreme' bottle neck might be placed to within perhaps 0·5 mm of its correct position under the filling head, a deviation which any filling machine should be able to cope with quite easily.

OVALITY

Ovality is a term used to describe how far a nominally cylindrical bottle departs from a perfectly circular cross-section. A diameter of, say, 50 mm with a tolerance of ± 1·1 mm normally means that a bottle with one horizontal dimension of 51·1 mm and the other of 48·9 mm would be acceptable; such a bottle would be said to have an ovality of 2·2 mm.

If the method of handling and packing will permit the amount of ovality implied by the diameter tolerances, the bottle maker will be able to make the best use of his moulds. As the mould cavity wears larger, he

D

can for a time restore it to its original volume by slightly shaving down the contacting faces of the two mould halves, thus giving a slightly oval shape.

For some bottles, especially those which are conveyed or labelled by rolling them on their sides, too much ovality could cause trouble. In such cases it might be necessary to put a further restriction on the ovality allowed. In such cases it is usually found that the ovality can be kept within 75% of the stated tolerances.

METHODS OF CHECKING DIMENSIONS

A specification of dimensional tolerances is not complete unless the method of gauging or measuring the dimensions is also specified. The required methods are usually obvious, provided that everyone concerned keeps clearly in mind that the purpose of checking dimensions is to ensure *that the bottle will perform satisfactorily and efficiently on the packer's filling line and throughout its subsequent career.* If this principle is lost sight of, there is sometimes a danger that the whole business of checking may degenerate into a sort of 'gaugemanship'. Practical methods for checking dimensions are discussed in Chapter 6.

Capacity

Probably the most important single property of a container is its capacity. Mostly capacities are specified in the metric units of the millilitre (ml) and the litre (l), although the intermediate centilitre (cl) is also permissible.

The old imperial units of the fluid ounce and the pint may, like old soldiers, never die, but they are certainly fading fast. The similarly named (but differently sized) American units are also progressively disappearing in favour of metric units.

Automatically made bottles may range in capacity from 5 ml or even less up to around 3 litres, and the range may be extended to 5 litres or more with semi-automatic processes.

CAPACITY DEFINITIONS

The capacity of a bottle is defined either as the brimful capacity or the capacity to a stated filling height. The former of these lends itself better to quick and accurate measurement, except for wide-mouth jars, and is generally more convenient for accurate capacity control during manu-facture. On the other hand, most filling machines are designed to fill to within a fixed distance from the top of the bottle, and so the packer may be more concerned with the capacity to this level than with the brimful measurement. (There are some exceptions to this in the machines used for filling carbonated liquids, where the volume of liquid delivered depends on the brimful capacity of the bottle and on the machine settings. See Chapter 8.) Either method may be used in capacity specification, and the bottle maker will control the capacity at whichever height is required. (He cannot control at both heights simultaneously.) The capacity tolerance will probably be the same in either case.

What exactly is meant by 'brimful'? Surprisingly no official definition has yet been given for the meaning of this as applied to bottles. The following definition is recommended as the most practical and satisfactory for all types of glass container:

'A bottle is brimful when it is standing on a level surface and is filled with water until the centre of the surface of the water is level with the top of the bottle.' (See Fig. 3.1.)

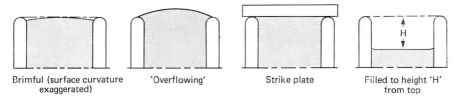

Brimful (surface curvature exaggerated) 'Overflowing' Strike plate Filled to height 'H' from top

Figure 3.1. Brimful and other filling conditions

The corresponding French term *ras-bord* or *ras-col* means literally 'level with the brim', and the Germans refer to *Rand voll* or *strichvoller Inhalt*, literally 'strike-full capacity'. The usual American term, however, is 'overflow capacity', which is misleading because they do *not* mean the condition in which the bottle is just starting to overflow. Such a condition would be unrealistic and very difficult to measure accurately, and in fact they mean exactly the same by 'overflow' as we do by 'brimful'.

An alternative definition of brimful is the condition obtained by using a 'strike plate' (Fig. 3.1). Strike plates are used when judging the fullness of legal measuring vessels, but if used on bottles they would give a capacity very slightly higher than that obtained according to the recommended definition. For narrow-mouth bottles the capacity difference between the two systems is quite negligible, and as the use of a strike plate would be an unnecessary complication, the definition has been worded accordingly. For jars with very wide mouths, some form of strike plate becomes essential for accurate measurement (Fig. 3.2), but nevertheless it is considered more satisfactory to stick to the same definition for all types of container, and to make allowances, if necessary, for the very slight discrepancy which occurs when a strike plate is used. (See section on capacity measurement in Chapter 6.)

May need slight greasing   Allowance needed for
                                            additional volume of plate

Figure 3.2.   Strike plates for capacity measurement of jars

The definition of capacity to a filling height at a stated distance from the top of the bottle is almost self-evident; the exact meaning is illustrated in Fig. 3.1.

Some types of filling machine fill not to a fixed distance from the top of the bottle but to a fixed distance upwards from the base plate on which the bottle stands. Theoretically the bottle capacities should be measured and specified in the same way, but in practice this would be another unnecessary complication. It might be necessary if bottle capacities depended to a significant extent on their overall heights, but with most types of bottle this is not so.

CAPACITY TOLERANCES

Modern methods of mould making have enabled the bottle maker to produce mould cavities to a volume accuracy within 0·1 or 0·2%, even when the shapes are complex. To obtain the highest precision, the moulds are regularly taken off the forming machine to be cleaned and checked for dimensions and capacity. The basic methods of controlling the volume of molten glass were discussed in Chapter 2; in addition it is becoming possible to weigh the hot bottles automatically as soon as they leave the machine, to provide additional control. The final part of the bottle-forming operation is putting the glass in the right place in the mould; the problems involved in this have already been discussed.

The degree of capacity control which is possible for any bottle depends on its size and shape, and on the effectiveness of the control measures which the bottle maker is able to apply. Plainly, it is not possible to guarantee the precise capacity of every bottle; nor is this necessary in practice. Past experience has resulted in the production by the Glass Manufacturers' Federation of the following table of working capacity tolerances, which like the dimensional ones are set at the 2-sigma level.

This table distinguishes between round (cylindrical) and non-round bottles, and between *normal* manufacturing control methods and *abnormal* methods. Normal control methods mean making regular checks on the capacities produced and making process adjustments to keep all the bottle capacities as close to the ideal as possible, whereas abnormal implies that the complete range of capacities produced is reduced by rejecting a significant proportion of the production.

In the table, column A relates to non-round bottles controlled by normal methods and column C to round bottles controlled by abnormal methods. Column B relates both to round bottles with normal control methods and non-round bottles with abnormal control methods.

TABLE 4: *Working Capacity Tolerances*

| Nominal capacity (ml) | Tolerances (ml) | | |
| | A | B | C |
Up to and including	±	±	±
25	1·7	1·3	1·0
50	2·5	1·9	1·4
75	3·1	2·3	1·7
100	3·6	2·7	2·0
125	4·0	3·0	2·2
150	4·4	3·3	2·5
175	4·7	3·5	2·7
200	5·1	3·8	2·8
250	5·6	4·2	3·2
300	6·1	4·6	3·5
350	6·7	5·0	3·7
400	7·1	5·3	4·0
450	7·6	5·7	4·2
500	8·0	6·0	4·5
600	8·7	6·5	4·9
700	9·5	7·1	5·3
800	10·1	7·6	5·7
900	10·7	8·0	6·0
1000	11·2	8·4	6·3
1250	16·7	12·5	9·4
1500	20·0	15·0	11·2
1750	23·3	17·5	13·1
2000	26·7	20·0	15·0

MEASURING CONTAINERS

Apart from the above recommendations, two EEC Directives were approved in December 1974 relating, among other things, to the capacities and tolerances of approved 'Measuring Containers'.* The tolerances for measuring containers are set out in Table 5; it will be seen that they correspond generally with those in columns A and B of Table 4, but the degree of correspondence varies at different points in the table because of the broader 'steps' chosen for the EEC specification.

TABLE 5: *Capacity Tolerances for Measuring Containers*

Nominal capacity C (ml)	Capacity tolerances % of C	ml
	±	±
50–100	–	3
100–200	3	–
200–300	–	6
300–500	2	–
500–1000	–	10
1000–5000	1	–

By comparing the tables, it may appear that some bottle shapes can be controlled to tolerances tighter than thoses pecified for measuring containers. This is partially true only, because although both sets of tolerances are related to the 2-sigma concept, it will be seen in Appendix C that the actual control test procedures for measuring containers are more complicated and stringent than those used in ordinary circumstances.

Quality control

The purpose of factory quality control is primarily to ensure that the bottles passed by the various inspection processes do meet the specified tolerances, and that all types of defective bottle have been removed effectively enough to ensure good performance. Also the quality controllers should provide a feed-back of information about quality to the forming machine operators, so that corrective action can be taken quickly

---

* For further information on EEC Directives and Measuring Containers, see Chapter 5, and Appendices B and C.

if needed. This requires a complex programme of monitoring and verification, both on the production line and in the testing laboratory.

As indicated in Chapter 2, bottle dimensions and glass weight are checked continually as the containers come off the forming machines, and a full inspection is carried out as they emerge from the annealing lehr. This inspection used to be carried out by a team of sorters stationed around the end of the lehr, but in most cases today the bottles are marshalled into a straight line before being presented to sorters stationed alongside the single line conveyor.

Once the containers are in a straight line, automatic inspection devices are used to supplement the manual sorting operation, and a great deal of development work is now going on in the field of automatic inspection.

Among the most common automatic inspection devices are those that rotate each bottle at high speed at a number of different indexing stations. Each station carries out a different mechanical or electronic check, such as gauging for maximum or minimum height. These machines (see Fig. 3.3) can also check for such faults as uneven sealing surfaces and cracks or splits in the glass. Automatic counters keep a tally of the defects found at each station so that any necessary remedial action can be taken at the

Figure 3.3. Automatic inspection

forming machines or elsewhere in the various manufacturing processes.

One drawback of the first automatic gauging and inspection machines was that they were complex and required considerable time and skill to set up. So until recently they were used only for checking long runs of bottles. However, later machines are now beginning to become more practicable for shorter runs.

After inspection and packing, statistical samples are taken to the laboratory and tested to confirm that the packed ware meets the required tolerances and complies with any other defined standards.

The dimensions of the samples are checked by measuring with a 'go, no-go gauge', usually set at limits tighter than the specified tolerances in order to ensure adequate compliance.

Capacity checking of samples is done in the usual manner, by weighing bottles first empty and then filled with cold water. The weight of water in grams can be converted to volume in millilitres by multiplying by a factor slightly greater than 1, the exact value depending on the water temperature. (Fuller details of this conversion are given in Chapter 6.)

Examination of the individual capacity results is usually sufficient to ensure that the bulk production will comply with the specified tolerance. If, however, the production consists of Measuring Containers, the quality control laboratory will probably carry out additional tests or calculations periodically to ensure full compliance with all the Measuring Container requirements.

Most packers have found that the quality control carried out by the glass manufacturer is adequate to ensure satisfactory performance, and no further checking is needed when the bottles are delivered. The filling line itself is the most reliable judge of quality. In some instances, however, users may choose to carry out their own acceptance testing of bottles, and this is discussed further in Chapter 6. It is important to remember though that quality can mean different things to different bottle users – even among those using exactly the same bottle. Some may place most emphasis on some aspect of functional performance, while others may be more concerned with the visual appearance of the pack. The glass container manufacturer therefore needs to know exactly what each customer expects in terms of quality so that the quality control programme in the glassworks can be adjusted to meet each set of needs. It is a fact of life in the glassworks that any process adjustment made to reduce or eliminate one type of defect often brings the risk of increasing the incidence of some other characteristic. So it is no solution to the quality problem just to tell the bottle maker to 'make them all perfect'.

4. Physical and chemical properties of bottles

The most important physical property of a bottle is of course its strength, that is its ability to withstand forces. The forces may be applied internally, usually by the pressure of the gas or liquid it contains, or externally, by impacts and applied loads or by the pressure of the external atmosphere when it is higher than the internal pressure. In addition internal forces may be induced thermally, because parts of the bottle at different temperatures will try to expand by different amounts.

Before discussing this in detail, we need to be quite clear about the meanings of the main terms used, such as pressure and strength, and the units in which they should now be measured:

Pressure is strictly the force acting on a surface when some substance, be it gas, liquid or solid, is pushing against the surface. In the past, we generally measured this pressure in pounds per square inch (psi), understanding that 1 psi was the force exerted by gravity on a mass of 1 pound resting on a surface area of 1 square inch. Scientifically this was not quite satisfactory, because the force of gravity is a slightly varying quantity. So we now use a unit of pressure which has an absolute meaning, and which can be related to the other metric units in general use. This is the *Bar*, which has been defined so that 1 bar is the pressure exerted by the earth's atmosphere at sea level under specified conditions. For most practical purposes 1 bar is equivalent to 14·50 psi, so whereas we used to refer to a pressure in a bottle of say 150 psi, we should now convert this to 10·34 bars.

Stress is the force acting on an area within a solid body, and if it is a *tensile* stress it will be a force trying to pull the body apart. For glass containers the only stresses of practical importance are tensile stresses acting close to the glass surface, so we are referring to forces parallel to the surface, acting on a small area at right angles to that surface. (These stresses can be produced by various means, as will be discussed later.)

Thus stress, like pressure, can be considered as a force per unit area, and it has often been expressed in psi. Any confusion so caused, however, has now been removed, because we should in future use the fundamental metric unit of force, the *Newton*. One newton per square metre is a very small stress so the practical unit chosen is the meganewton per square metre, MN/m^2, which in practice is equal to 145 psi. Thus a tensile stress expressed as say 5000 psi now becomes 34·5 MN/m^2.

It may be noted that 1 MN/m^2 = 10 bars exactly, but while it would be incorrect to express stresses in bars, it is permissible to quote pressure in kilonewtons or meganewtons per square metre. So for example a pressure of 150 psi should be expressed as 10·34 bars, or alternatively as 1·034 MN/m^2 or even 1,034 kN/m^2.

Strength is simply a statement of the maximum pressure or other imposed force, or alternatively the maximum stress caused by an imposed force, which the article is expected to withstand without failing. Thus we might say that a bottle has an internal pressure strength of 12 bars, or that the tensile strength of its surface is 30 MN/m^2. But in the latter case, we should bear in mind that the tensile strength of a bottle surface may not be the same at all points of the bottle, either because it was different initially, or because it has been treated differently. So when speaking of a bottle's tensile strength, we should if possible specify the part of the bottle and its condition, e.g. outer sidewall after filling-line abrasion.

STRENGTH MAINTENANCE

In order to ensure the best possible performance of bottles in service, it is necessary to minimise the stresses incurred, firstly by good design and manufacture of the bottle and by good design and operation of the subsequent packaging operations; and secondly by maximising the bottle's ability to withstand those stresses. Dealing with the second point, the whole new science of glass surface treatment has done a great deal to help retain adequate surface strength throughout the bottle's life.

As indicated in Chapter 1, glass is probably some twenty times stronger than steel when initially formed, but it rapidly loses much of this strength because of the development of sub-microscopic surface flaws. As a result of these flaws, the strength of an ordinary bottle outer surface when it goes into service has already fallen from 750 to 75 MN/m^2, and this may be reduced even more during the early part of its life due to the effect of impacts and abrasions.

In the early 1960s it was found that if a hot glass surface is treated immediately after forming with certain organic compounds of titanium or tin, a reaction takes place which changes the nature of the glass surface and makes it more resistant to the formation of weakening surface flaws.

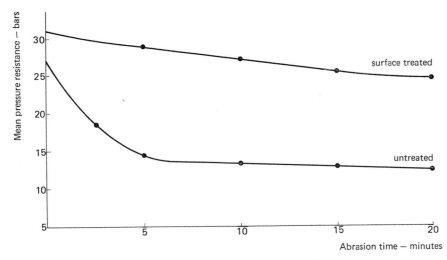

Figure 4.1. Effect of simulated filling line abrasion on internal pressure resistance of 0·75 litre bottles

Since then, this basic 'hot end' treatment has been considerably improved and refined, and has been supplemented by new 'cold end' treatments whereby organic lubricants or other materials are applied to the bottles after cooling. The methods used are outlined in Chapter 2. Besides improving the initial strength of the bottles, these new techniques improve the 'lubricity' of the containers, helping them to move smoothly against each other and minimising the weakening effects of abrasion. Fig. 4.1 shows how the strength of a typical bottle can be enhanced throughout its service by good surface treatment.

INTERNAL PRESSURE RESISTANCE

Of the various factors which need to be considered in relation to mechanical strength, internal pressure resistance (or 'bursting pressure') is the most straightforward subject, for two reasons: because it can be measured accurately, and because there is a clear correlation between pressure resistance and the main features of design. Other aspects of strength are more complex, and less amenable to quantitative treatment. Fortunately features of design which give good pressure resistance are also good for the other types of strength. Therefore it is sensible to consider pressure resistance first, even though not every bottle has to withstand internal pressure.

The pressure resistance or bursting pressure of a bottle is measured by applying an increasing internal pressure, either steadily or in a series of steps, until the bottle breaks. The results obtained in any test depend to a considerable extent on the conditions in which it is carried out; we cannot discuss pressure strength in a precise way until we have considered the variability in the test methods, and decided on a standardised test procedure. There are two factors which are responsible for this variability:

1. *Rate of pressure rise.* Bursting pressures depend on the rate at which pressure is applied. If a steady pressure is applied to a bottle one of three things may happen: it may fail immediately, it may survive indefinitely, or it may fail after an interval of time. The reason for the third possibility is that there may be a very tiny mark or flaw on the surface at the point of eventual failure, which gradually grows under the applied pressure until finally the bottle breaks suddenly. So if the pressure is increased slowly, or in steps with long time intervals between, the flaws will have more time to develop, and the bursting pressures will be reduced.

Different workers* have found much the same effect, and Fig. 4.2 summarises the most comprehensive investigations.

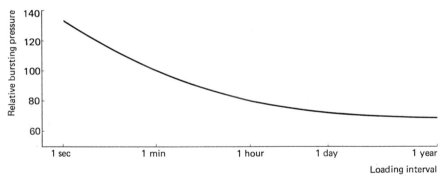

Figure 4.2.   Effect of loading interval on bursting pressure (logarithmic time scale)

In this diagram the values obtained with a 60-second interval have been regarded as the 'standard' bursting pressure and given an arbitrary value of 100. For practical purposes, we may deduce that if a pressure is to be maintained for a time of the order of 15 minutes to 1 hour (as in the pasteurisation of beers) the effective resistance of the bottle will be about 80% of the standard value. If the pressure is maintained for one day or more, up to at least one year (as in the storage of carbonated liquids), the effective resistance is about 70% of the standard value. For pressures maintained for only a few seconds, the effective resistance might be 130% of the standard value, but this is of no practical importance to the packer.

* F. W. Preston, *Bull. Amer. Cer. Soc.*, 1939, **18**, 35; T. C. Baker and F. W. Preston, *J. App. Phys.*, 1946, **17**, 170 and L. H. Lehnert, *Glastech. Ber.*, 1955, **28**, 380.

2. *Effect of temperature.* There has been some disagreement in the published work as to whether this effect exists, but the author has obtained
confirmation in laboratory measurements that the strength of glass does
decrease slightly as the temperature rises. Results obtained on lightweight
beer bottles with a weight:capacity ratio of about 0·5* were as follows:

Temp. (°C)	Mean bursting pressure (bars)	Lowest bursting pressure (bars)
10·5	27·6	16·2
60	23·4	16·2
98	21·0	14·8

Thus the mean bursting pressure of this bottle at 98°C was about 75%
of its value at room temperature. The minimum bursting pressure was,
however, reduced by less than 10% over the same temperature range.

(There are several effects, especially that of surface abrasion, which alter
bursting pressure distributions in this same way, i.e. they appear to
weaken the strongest bottles much more than they do the weaker ones;
this is discussed in more detail in the next section.)

MEASUREMENT OF PRESSURE RESISTANCE

In most pressure tests, the bottle is filled with water, and the pressure
applied by a pump or a direct ram. With manually controlled apparatus,
the pressure should be raised in steps of about two bars, at intervals which
can be adhered to accurately, say 30 seconds. Automatically controlled
apparatus can work with smaller time intervals, down to three seconds.

For specification or comparison of pressure resistance, it is customary in
all countries to assume standard test conditions of 60-second intervals,
applied at room temperature, and if the actual test interval differs from
this it should be allowed for by the following factors:

Test interval (seconds)	Multiply test pressure by
1	0·75
3	0·81
10	0·89
20	0·93
30	0·96
60	1·00

* This ratio is obtained by dividing the weight in grams by the brimful capacity in
millilitres.

Figure 4.3. Automatic pressure tester

Most mechanised incremental pressure testers make this correction automatically. The most modern automatic tester, shown in Fig. 4.3, has dispensed with intervals altogether but raises the pressure continuously at a fixed rate; the pressure shown on the digital read-out has already been adjusted internally to be equivalent to a 60-second interval test.

VARIATION IN PRESSURE RESISTANCE

When a bottle maker is asked to make bottles which under standard test conditions have a bursting pressure of, say, not less than 12 bars, the user may be surprised to find that some of them have much higher resistance, of perhaps up to 25 bars. In fact if a large group of bottles are tested, the results will usually form a Gaussian distribution, similar to that obtained with dimensional variations, and discussed in Appendix D.

For bottles of average weight the ratio between the highest and lowest results is generally about 2:1. For very lightweight bottles, the spread might be rather less.

The reason for this spread is that the bursting pressures depend very critically on the details of the glass distribution, and on the surface condition of the glass, i.e. how much and how severely it has been rubbed or abraded before testing. There is another reason for variation, in the submicroscopic flaws which exist to varying extents on different glass surfaces. As indicated earlier, the effect of these flaws can nowadays be limited to some extent by the surface treatments described in Chapter 2.

While discussing the spread of bursting pressures, we should mention the tantalising fact that if tests are carried out on brand-new bottles before they have been touched by hand, paper, cardboard or any other material, they show an abnormally high strength. At the other end of the scale, after bottles have been in prolonged use, the strength will, not surprisingly, have been reduced. The general pattern of pressure strength in the life of a bottle may be illustrated by the following data; the figures are hypothetical, but correspond to what might be expected from a reusable bottle:

Condition	Brand-new and untouched	New condition as first delivered to user	After prolonged use
Bursting pressure (bars)			
Highest	60	33	20
Mean	40	23	15
Lowest	25	13	10

It is a curious fact that the effect of prolonged wear is mainly to weaken the strongest bottles, while the bottles which were originally the weakest seem to retain their strength much better. This is undoubtedly part of the reason why it is possible to get such long and safe service out of well-designed bottles. Even after as much as ten years continuous service, well-designed beer and mineral bottles have been found to have adequate strength left.

EFFECT OF DESIGN ON PRESSURE RESISTANCE

It is usual to start a discussion on pressure resistance of bottles by saying that the ideal shape is a sphere. This is true in a very limited sense, but every packer will know that a spherical bottle would be a nightmare to handle on a modern filling line, and we must dismiss it as quite impractical. The shape which is next highest in theoretical pressure resistance is the cylinder, and fortunately this shape has everything to commend it from other practical points of view.

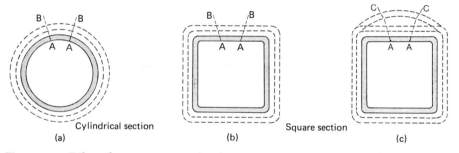

Figure 4.4. Effect of pressure on round and square cross-sections (amount of expansion greatly exaggerated)

The reason for the superiority of the cylindrical shape may be seen from Fig. 4.4. When a uniform cylinder is subjected to internal pressure, it expands uniformly as shown in 4.4.a, until the piece marked AA reaches the position BB. Every piece of the cylinder is stretched by the same amount; in other words the stress in the surface is the same at all points, and the amount of tension can easily be calculated. When any less symmetrical shape is under pressure, two different effects occur. The first effect, shown in 4.4.b, is the general increase in size due to the pressure; the tension produced is practically the same as for the cylinder. But in addition some distortion of the original shape occurs because some parts of the vessel are better able to resist the pressure than others. The parts which move furthest are the centres of flat panels, and as shown in 4.4.c these are the parts which are stretched most, to position CC in the diagram. So the total stress is made up of two parts, the uniform expansion (AA to BB) plus the non-uniform part (BB to CC) caused by the distortion or bending effect. In practice the first part is usually small: the change in dimensions of a cylindrical bottle under pressure can hardly be measured; but the second part may be large: the distortion of a large asymmetrical bottle under pressure might easily rise to 0·1% of the original dimension. (The distortion is not invariably in an outward direction; for a bottle of rectangular section the larger sides may bow outwards under pressure while the smaller sides curve inwards.)

The stresses produced in a square bottle under a given pressure will thus be much higher than the stresses in an equivalent cylindrical bottle under the same pressure; so if the pressure is increased in both, the square bottle will reach breaking point at a lower pressure than the cylindrical one.

THEORETICAL CALCULATIONS OF PRESSURE RESISTANCE

Theoretical calculations are not able to predict exact values for bursting pressures of bottles, but they are nevertheless useful to illustrate the effect of the main design features, so we will pursue them a little way.

When a thin-walled hollow cylinder, of internal diameter d, and wall thickness t, is subjected to an internal pressure p, the tensile stress at every point in the outer surface s, is given by

$$s = \frac{pd}{2t}$$ (assuming that s and p, also d and t are measured in the same units)

Failure is liable to occur when the stress in the outer surface reaches its limiting values, which for bottle glass is of the order of 35 MN/m²,* so we may say that the bursting pressure of the cylinder, p_{max}, is given by

$$p_{max} = 10 \times 35 \times \frac{2t}{d} \quad \text{(bars)}$$

Immediately we can deduce that pressure resistance is proportional to wall thickness, and inversely proportional to the bottle diameter; both conclusions are approximately true in practice. To examine the actual values, let us consider a bottle of wall thickness 1·5 mm and diameter 70 mm. Theory gives:

$$p_{max} = 350 \times \frac{3}{70} = 15 \text{ bars}$$

* This value would be reasonable for new bottle glass after going through the usual sorting, packing and transport processes. Before these processes, it might temporarily be as high as 75 MN/m² or more.

E

This value is quite realistic, but there are several other factors which will affect the strengths obtained in practice:

1. The bottle shoulders and base usually have a stiffening effect, if the cylinder is not too tall in relation to its diameter.

2. In practice the bottle will not be perfectly cylindrical, and it will not have a perfectly uniform wall thickness. Theoretical and practical investigations suggest that the amount of ovality which is usual in a bottle could reduce the theoretical bursting pressure by perhaps 25%. With regard to wall thickness, more accurate predictions can be made if the *minimum* wall thickness is used in the equation.

3. We have not allowed for the compressive stresses left in the bottle surface by the annealing process, which will help the bottle to resist internal pressure.

When we come to attempt theoretical calculations for shapes other than cylinders, the 'errors' become so large that the results are of little quantitative value. The general equation for the maximum stress in a non-cylindrical tube may be written:

$$ s = p\left[K_1\frac{d}{2t} + K_2\left(\frac{d}{2t}\right)^2 \right] $$

where K_1 and K_2 are factors depending on the particular geometry (the two terms inside the bracket correspond to the two separate pressure effects shown in Fig. 4.4.b and c, so for a cylinder $K_1 = 1$ and $K_2 = 0$). For the full treatment of this problem, the reader is referred elsewhere.[*] Here we shall only point out that the second term in the equation becomes relatively large for most non-symmetrical shapes, for which K_2 is large, because of the term $\left(\dfrac{d}{2t}\right)^2$. The bursting pressures calculated for bottles with elliptical, rectangular and similar cross-sections are, however, unrealistically low. For example, theory suggests that an elliptical section with the major axis twice the minor axis is 50 times weaker than a cylinder; whereas in practice bottles of this shape are only about twice as weak as cylindrical ones.

For practical purposes it is much better to use the ratios published by F. W. Preston;[†] these have been found to be a very useful practical guide, and no amendments have been necessary:

[*] For example W. M. Hampton, *J. Soc. Glass Tech.*, 1941, **25**, 121.
[†] F. W. Preston, *Glass Ind.*, 1940, **21**, 272.

Shape	Ratio of pressure resistance
Cylinder	10
Ellipse (major axis twice the minor)	5
Square with 'well-rounded corners'	2·5
Square with sharp corners	1
To this we may add:	
Rectangle, with length equal to twice width, and with rounded corners	1

The rounding of sharp edges is evidently important, and here the theoretical studies of W. M. Hampton can give some quantitative guidance.

Considering a square cross-section, of side L and corners with radius of curvature, C, the theoretical bursting strength will vary as shown below, for different wall thickness (t):

C/L	$t=L/10$	$t=L/20$	$t=L/30$
	Bursting pressure (bars)		
0·0	11·5	3·1	1·4
0·1	12·6	3·5	1·6
0·2	14·6	4·1	1·9
0·3	18·6	5·3	2·5
0·4	27·6	8·6	4·2
0·5	69·0	34·5	23·0

We have not discussed the shape of the shoulders and neck or the base, because it plays little direct part in determining the pressure resistance. Almost any shape of shoulder and base will provide approximately the same amount of stiffening to the side walls. It is nevertheless important that the shoulder should be reasonably streamlined and the base joined to the side wall by a smooth curve ('insweep'), for three reasons:

1. A smooth streamlined contour is a great help to obtaining good glass distribution, and this helps pressure resistance directly.

2. If part of the shoulder becomes too flat there is a danger that it may become less strong in pressure resistance than the side wall.

3. If either base or shoulder has sharp edges, it might be unduly prone to damage by impact, and this could lower the pressure resistance.

PRESSURE RESISTANCE IN PRACTICE

It is evident that by varying the glass thickness, the design and the surface condition it should be possible to obtain almost any pressure strength which might be required. This is technically true, but in practically all the markets where bottles contain pressurised products it is also important to minimise packaging costs, so the real problem is how to ensure adequate strength most economically.

The first step towards achieving this has been to design all such bottles with round cross-sections at all heights, and in most cases they approximate closely to the 'ideal' cylinder with streamlined top and bottom. The other thing which can be done to ensure sufficient glass thickness at lowest cost is to keep the bottle height as small as possible in relation to the diameter. Obviously the greater the ratio of height to diameter, the greater will be the surface area of the bottle required to hold a certain volume, and so the greater will be the weight of glass required. Not many designs are short enough to be ideal in this respect, but the further they depart from the ideal shape the more their weight needs to be increased.

The question remains as to what minimum bursting pressure should be specified in practice. As will be seen from Appendix F, the pressure generated by carbonated liquids rises with temperature, either when the product is heat-treated, as in the pasteurisation of beer, or when the product is warmed up by the external environment during transport or storage. In British conditions, the highest internal pressure which can be reasonably anticipated with highly carbonated drinks is in the region of 5 to 6 bars. A 'safety factor' of about 2 has been found empirically to give a satisfactory balance between strength and cost, so it has become usual in the UK to specify a minimum pressure strength of about 10 bars (145 psi). All the British Standards so far published (see Chapter 5) have specified 10·34 bars, at the point in time when the new bottle is first received by the packer, and our evidence to date is that this is satisfactory.

Bottle strength and product safety is of course an evocative subject, and there has been a tendency with some products and in some countries to ask for higher strengths, usually on the general grounds that the more severe the specification, the better should be the bottle performance. But in British experience this seems to be largely a fallacy: if some bottles fail prematurely, for example on the filling line, it is invariably found that they were damaged or defective, not that their strength specification was too low. So the right way for the glass maker and the packer to ensure best performance is to concentrate on preventing bottles being made defective and allowed to go into service. Aiming for a higher strength specification could sometimes help, but it would always have the main effect of leading to a heavier and more costly bottle.

When comparing strength specifications and proposals from different parts of the world, it will be found that there is a wide divergence of pressure figures, ranging from 8 to 27 bars. It should be borne in mind that these relate to different designs of bottle, different products, conditions of use and environmental conditions, and most importantly different types and times of test, ranging from tests during glass manufacture to tests on bottles after a period of service.

VERTICAL LOAD STRENGTH

Glass can resist very severe compression, and if external forces are applied to a bottle in such a way that only compression is caused, then damage or breakage is almost impossible. For example, when a capping machine presses down fairly steadily with a weight of, say, 150 kg on the top of a well-designed beverage bottle, all parts of the shoulder and sides will be evenly compressed, and no significant tension will be produced. If, however, the force is applied in such a way, or the shape of the bottle is so contrived, that some parts of the bottle surface are severely stretched, breakage will occur if the tension anywhere reaches the limiting value, probably around 35 MN/m² for the outer surface.

In most designs the greatest stress occurs in the region of the bottle shoulder. Fig. 4.5 shows the average bearable load for four shapes of shoulder, and illustrates graphically the relation between radius of vertical shoulder curvature and mean bearable load.

This graph is only an approximation to the facts, because it does not

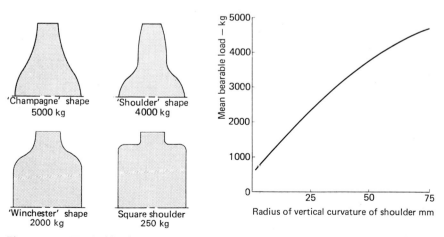

Figure 4.5. Vertical load resistance of different shoulder shapes

take into account wall thickness or body diameter. In fact, wall thickness is surprisingly unimportant with the stronger shapes; for example, light-weight non-returnable bottles have just as much vertical load strength as heavier reusable bottles, if both have the same 'champagne' shape shoulder. The width or diameter of bottle is more important; the most accurate criterion of load strength would be the ratio of the radius of vertical curvature to the radius of horizontal curvature (i.e. the body radius), but the simplified version in Fig. 4.5 should be adequate for any design needs.

In a few designs the shoulder is stronger than other parts of the bottle, for example the region just below the finish, if there is an outward curve in the neck, or the joint of the base and side wall. In any bottle design it is desirable to keep all radii of vertical curvature as large as possible, if maximum load strength is required.

Crushing strengths may be measured in any suitable form of hydraulic press. For accurate measurements, the same principles should be followed as are used in bursting pressure measurements; for example, the load should be increased in steps at constant intervals of time.

IMPACT STRENGTH

When a bottle receives an impact, the glass at the point of contact will be compressed, but the surrounding zone will be put in tension, and there will also be a region of tension on the inside surface below the point of impact. Also shock waves travelling through the glass may occasionally produce tension some distance away.

There are basically two sorts of impact: those in which a moving object strikes a stationary bottle, and vice versa, as for example when a bottle is dropped on the ground. In the first case the outcome depends on the sharpness and hardness of the striking object, and on its kinetic energy, i.e. on its weight and speed. It also depends on whether the bottle is free to move away when struck, or whether it is partly or completely res-trained in position. If it is filled with something heavy, its freedom to move away quickly is restricted.

In the second case the outcome depends on how rigid and immovable the obstacle is. If the energy of movement of the bottle is destroyed almost instantaneously by meeting a very rigid obstacle, the stress pro-duced in the bottle will be high. If the bottle is filled with liquid there will be an additional effect, as the momentum of the liquid will be converted into a momentary internal pressure, which can easily rise to 25 bars or more. As before, the extent of this pressure rise depends upon how rapidly the bottle is brought to a halt. This explains one of the basic purposes of providing cushioning around and under a filled bottle when it is being transported: namely to ensure that any deceleration of the bottle is sufficiently controlled to prevent the internal pressure rising to a dangerous level. Likewise it is important that any bottle containing liquid should have reasonably good pressure resistance, in order to help it withstand this hazard.

It is not necessary, nor would it be possible, to specify exactly what severities of impact a particular bottle should be able to withstand. If there are unavoidable reasons for a bottle receiving heavy impacts, then the designer will take this into account in his plans. (Nearly always, however, those reasons turn out on investigation to be anything but unavoidable.) The principles employed to obtain greatest impact resistance are exactly the same as for internal pressure resistance and vertical load resistance.

IMPACT TESTS

Measurement of resistance to impact is not an easy subject. The conventional approach is to use a rigid pendulum with a hardened steel ball as the 'bob', the vertical height through which the striker falls being a measure of the impact energy. Each bottle is struck in exactly the same position. This method is useful for research and for design testing, but it tends to give very detailed information about one point on the bottle, without necessarily being a good guide to its general performance in service.

The second type of impact, in which bottle and contents are moving, is probably the more important of the two. A simple realistic way of obtaining these conditions is to fill the bottle with water, seal it and drop it on a rigid surface. It is not usually necessary to make the bottles fall at all possible angles, and sufficient information can be obtained by dropping them either on to their bases, or flat on their sides. This kind of test can be made progressive by testing every bottle with a series of drops from increasing heights until it breaks, and then calculating the mean height required to produce breakage.

RESISTANCE TO ABRASION AND SCRATCHING

The initial surface strength of a bottle depends on the glass itself and on the standard of its surface treatment rather than on the bottle design. But the amount of strength lost in service may depend significantly on aspects of the bottle design as well as the surface treatment.

For example a new, well surface-treated bottle may have a strength over its outer surface of, say, 50 MN/m². During its service, some parts of the bottle may become abraded and their strength fall to, say, 40 MN/m² (or further without surface treatment). As we have seen earlier, when a bottle is subjected to internal pressure or other constraint, the stresses produced will be different in different parts of the bottle, and the bottle design should be adjusted if possible to flatten out and lower the peak stresses.

The second thing to do if possible is to adjust the design and/or the handling systems so that the peak stress areas do not coincide with the areas which receive most abrasion. Thus with bottles for pressurised liquids it is beneficial to design slight bulges at the top and bottom of the cylindrical part of the body; these will then receive the main abrasion on a conveyor, while the peak stress caused by pressure occurs on the protected part of the side wall between the bulges. Similarly it can be very helpful to further protect glass surfaces (including the side-wall bulges) by covering them with a pattern of small raised pimples or bars, called 'stippling'. If these are correctly designed, abrasion will be confined to the tops of the stipples, where the stresses are likely to be lower than elsewhere.

An apparently opposite technique may be effective sometimes. If a jar is required to have a very angular shoulder or a very 'square' corner between base and side wall, this will make those areas very sensitive to impact stresses, meaning that impacts at these points will cause greater stresses at the angles than elsewhere. So it could be wise to reduce the risk of such impacts occurring on a conveyor, by slightly tapering the jar at top and bottom, and transferring the contact point to a less sensitive area. (But, as discussed in Chapter 5, it would be better first to find out whether the angular shoulder was really necessary anyway.)

Thermal shock resistance

The words 'thermal shock resistance' mean exactly what they say, i.e. the resistance to sudden temperature changes. Some publications use the term thermal endurance for this property, but this is misleading, as the word 'endurance' implies resistance over a long period of time. Thermal endurance has been used in other contexts to define the extreme temperature a material will withstand before melting or burning, but as a glass container will withstand all temperatures up to 500°C this is a property of no practical concern to glass packers. In this book 'thermal shock resistance' will be used throughout.

THEORETICAL ASPECTS

If a glass article is at a uniform temperature and then one surface is suddenly chilled, the cooled glass will try to contract, but it is prevented from contracting freely by the warmer glass to which it is firmly attached. Consequently the cold surface is put into a state of tension, while the hotter surface is compressed. This is illustrated in Fig. 4.6, which treats a slab of glass as if it were made of two layers.

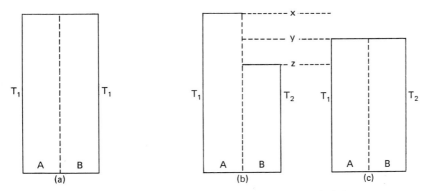

Figure 4.6.   Theoretical behaviour of a slab when one surface is chilled

Fig. 4.6.a shows the end view of the slab when both sides are at the same temperature $T.$ Side B is then cooled to a lower temperature, T_2, and if it were independent of side A, it would contract to the position shown in 4.6.b. But as it is attached to side A, it can only contract to the position shown in c, taking side A with it. Thus side A is compressed by an amount corresponding to the distance xy, and side B is stretched (i.e. put in tension) by an amount corresponding to yz.

This is an over-simplification of what actually happens: in practice that will be a progressive change across the section, from tension to compression, depending on how fast the heat flows through the glass. Also the stresses will be affected by whether the piece of glass is able to relieve some of the stresses by bending.

Calculation of the amount of stress produced by thermal shock is difficult even for symmetrical shapes, but many theoretical and practical studies have been made. The following equation indicates the relation which is found to exist between stress and temperature, when one surface of a hollow cylinder of soda-lime glass is suddenly chilled:

$$s = 0.342 \, (T_1 - T_2)\sqrt{t}$$

where s is the tensile stress in the chilled surface, measured in MN/m²,

T_1 is the initial uniform temperature of the article (°C),
T_2 is the lower temperature suddenly imposed on one surface (°C),
t is the thickness in mm.

Thus, for example, if $T_1 - T_2 = 30°C$, and $t = 4$ mm,

$$s = 0 \cdot 342 \times 30 \times 2\text{MN/m}^2.$$

If the effective tensile strength of the outside surface of the glass is taken as 30 MN/m² the relation may be expressed another way:

$$T_{max} = \frac{30}{0 \cdot 342\sqrt{t}} = \frac{88}{\sqrt{t}}$$

where T_{max} is the maximum temperature difference which the article can withstand, when the outside is chilled. Thus, for a hollow cylinder of wall thickness 4 mm, T_{max} would be expected to be 44°C.

If the sudden temperature change is in an upward direction, stresses will be caused in the same way, the compression being in the heated side and the tension in the unheated side. It will be clear that if both surfaces are heated simultaneously, no tension will be caused at either surface.

It is found in practice that the amount of tension produced in one surface of a bottle by suddenly chilling it is about twice as great as the

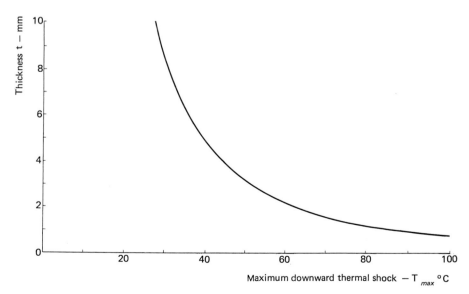

Figure 4.7. Theoretical relation between wall thickness and maximum downward thermal shock

tension produced by suddenly heating the other surface (assuming the same temperature change in both cases). In other words, glass articles can withstand twice as large a temperature jump in an upward direction as they can in a downward direction.

Wall thickness is an important factor because the thinner the glass the more difficult it is to establish a large temperature difference between the surfaces. The theoretical relation between wall thickness and maximum downward thermal shock is illustrated in Fig. 4.7.

THERMAL SHOCK TESTS

Theoretical investigations cannot cope with the complex shapes of bottles, or the effect of variations in wall thickness, but there is plenty of practical information available, from the routine thermal shock tests which are carried out during bottle manufacture. We can usually dismiss the possibility of breakage occurring on the inside surface of a bottle, because the effective strength of this surface is greater than 35 MN/m^2 (provided it is undamaged) and it can easily withstand a temperature drop of over $100°C$, at any likely wall thickness. So thermal shock tests are designed to test the weaker outside surface, and the most effective way of doing this is to stabilise the temperature of the whole bottle at one temperature, and then to chill the outer surface by immersion in colder water. The stabilisation can conveniently be achieved by totally immersing the bottle, in an upright position, in hot water for at least five minutes. It is then lifted out and lowered, still upright and therefore still full of hot water, into colder water, until the neck is nearly but not quite submerged. Automatic apparatus is generally used for routine testing, but it is possible to employ simple manual methods, provided that the water temperatures are carefully controlled (continuous stirring is essential) and that the time of transfer is kept constant.

Bottle shape has an important effect on thermal resistance; when the outside surface is chilled uniformly, the stresses are always highest near the join between base and side wall. Consequently, if thermal breakage occurs, it nearly always starts in this region. The reason for this lies in the bending effects of the stresses. In designing for good thermal shock resistance it is therefore important to ensure that the join between the base and side wall is not too abrupt; nearly all modern bottle designs provide a gentle curvature or insweep at this point.

Most modern designs of bottle when tested in the way described will withstand a downward thermal shock of around $40-45°C$. (From the theoretical equation we predicted that a hollow glass cylinder of wall thickness 4 mm and outer surface strength 30 MN/m^2 would withstand

44°C, so the agreement between simple theory and complex practice is remarkably good.) Very large or complex designs of bottle will, of course, not attain these levels, while small ones, particularly if very thin-walled and streamlined, might withstand up to about 70°C.

Optical properties

Completely colourless glass would transmit all wavelengths of visible light without any absorption. Commercial 'white flint' glass absorbs a very small proportion of the light which passes through it, but if it absorbs all wavelengths equally, it still appears to be colourless. This slight amount of absorption is usually too small to be measurable, and in controlling the colourlessness of white flint glass, the bottle maker has to rely mainly on the human eye, which is much more sensitive and discriminating than any optical instrument.

Transparent glasses appear coloured if they contain elements (added intentionally or as impurities) which absorb one colour of light more than another. For example, the yellowish colour of amber glass occurs because one constituent absorbs most of the blue and green light, and another constituent some of the red light from the spectrum, while the yellow light is transmitted.

If a coloured glass is selected simply for its looks, the usual procedure is for the bottle maker to supply sample bottles in the chosen colour, which can then be used for future reference or comparison. It should be remembered that the bottle maker has to maintain a constant composition and constant glass colour while making bottles of different sizes and shapes. The apparent colour of a bottle depends on the glass thickness (if the thickness is doubled, the absorption of light is quadrupled) and also partly on the bottle shape; bottles of different shapes and sizes may thus appear to have somewhat different colours, even though made of the same glass. Sample bottles should therefore be used only in assessing the colour of bottles of the same size and shape as the sample.

MEASUREMENT OF COLOUR

If an absolute measurement of colour is needed, the optical transmission curve of the glass can be measured by means of a spectrophotometer. This instrument passes a beam of light of any required wavelength through a specimen and records what percentage of the light is transmitted through the thickness of the specimen. Ideally, the specimen should be a flat polished plate with accurately parallel sides, but reasonable results can be obtained with a piece of glass cut from a flat part of a bottle. The spectro-photometer readings will depend on the glass thickness, and it is con-venient for comparison purposes to express all results as percentage transmissions for a standard thickness, usually 2 mm.

If C_1 is the percentage transmission for a thickness of bottle glass t_1, the percentage transmission for a thickness t_2 is given by C_2 in the equa-tion:

$$\log\left(\frac{C_2}{92}\right) = \frac{t_1}{t_2}\log\left(\frac{C_1}{92}\right)$$

This equation takes into account that 4% of the incident light is lost by reflection at each glass–air interface.

Typical transmission curves are shown in Figs. 4.8.a to d. Wavelengths are given in nanometres (nm)* and the visible spectrum extends from about 400 to 700 nm.

SPECIFICATION OF COLOUR

With modern methods of glass composition control and furnace control there is very little variation in colour. Specifications which fix the limits

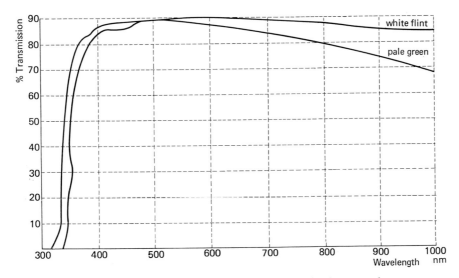

Figure 4.8a. Typical transmission curves for white flint and pale green glass, 2 mm thick stated in 4.8c)

* A nanometre is 10^{-9} metre. This unit has the same value as the millimicron which was the more commonly used unit in the past.

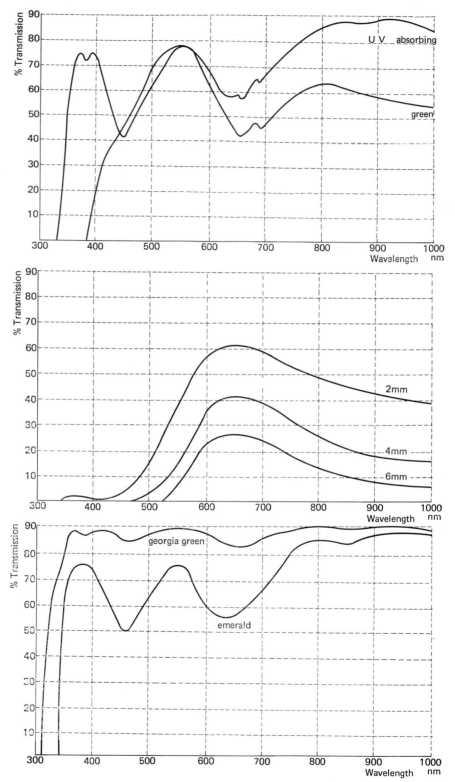

Figure 4.8b. Normal green glass and UV-absorbing green glass, 2 mm thick
Figure 4.8c. Amber glass, showing effect of different thicknesses
Figure 4.8d. American Georgia and emerald glass, 2mm thick

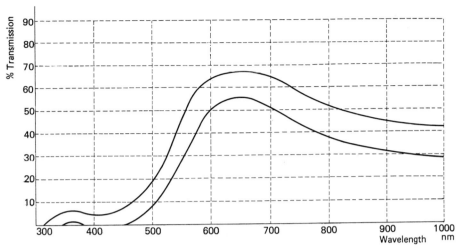

Figure 4.9. Limits of variation in amber glass production

within which the colour is allowed to vary are therefore rarely necessary. A possible exception to this is amber glass, which is somewhat sensitive to furnace conditions, and an example of the degree of control which is possible when operating a furnace over a long period is shown in Fig. 4.9.

Specifications have sometimes been set for amber glass in terms of the range of transmission permitted at a wavelength of 550 nm. This is not quite a complete specification of the colour, as it would theoretically be possible (but in practice very difficult) to alter the transmission at other wavelengths without affecting the value of 550 nm. If it is necessary to have a complete and absolute specification of colour, use may be made of the complex system of trichromatic coefficients, established by the International Committee on Illumination. A brief account of the system is given here for the sake of completeness; full details and methods of computation may be found in A. C. Hardy's *Handbook of Colorimetry* (Massachusetts Institute of Technology, 1936).

The basis of the ICI system is that any colour can be built up by mixing together the right amounts of standard white light and coloured light of a specified wavelength. The wavelength of coloured light required is called the 'dominant wavelength', and the 'purity' indicates the percentage of coloured light required (a purity of zero means 'perfect' whiteness*). The 'brightness' is a measure of the total amount of light of all wavelengths which reaches the eye.

Approximate data for the more important British colours are listed overleaf. They have been calculated for a thickness of 2 mm, except for

* The system was established in 1931, before synthetic detergents were invented.

pale green glass, which requires a thickness of 1 cm to give meaningful values of purity and brightness.

	Dominant wavelength	Purity	Brightness
	nm	%	%
Pale green	510	2	80
Amber	580	90	40
Green	560	20	70
Blue	470	40	35

LIGHT PROTECTION OF BOTTLE CONTENTS

It is well known that some chemical reactions can be started or accelerated when light of a particular wavelength falls on the materials concerned. The wavelength required depends on the particular reaction, but it is usually in the visible blue or invisible ultra-violet part of the spectrum, because the longer wavelengths in the yellow and red part of the spectrum do not usually possess sufficient energy. A good example of a photo-chemical reaction is human sunburn, which is activated by a wavelength of 300 nm. The fact that sunburn is uncommon in most parts of Britain shows that ultra-violet light is usually missing from British daylight, having been absorbed or scattered by the clouds and smoke.

If there is reason to think that light could harm a product, there may be several colours of glass which will provide protection. Ideally one should determine which wavelengths need to be excluded – the bottle maker should be able to assist with advice or with practical measurements – but even if this information is lacking, the use of amber glass is more than likely to provide adequate protection. From Fig. 4.8.c it will be seen that normal amber glass 2 mm thick excludes practically all light of wavelengths less than 450 nm, and for thickness of 4 and 6 mm, the 'cut-off' increases to 470 and 500 nm.

Green glasses which derive their colour from chromium oxide can also provide good protection, if they are melted in such a way that the chromium is in the correct chemical form; a transmission curve such as that shown in Fig. 4.8.b can then be obtained. Such green glasses tend to have a yellowish tinge, because the blue light is missing, and they are not much favoured by British packers. One American green glass, patented under the name 'Ultrasorb', has a visual colour barely distinguishable from the ordinary emerald colour. In some other American green glasses the ellowish tendency is allowed to develop fully, producing an unusual colour halfway between green and yellow (a colour which the French call *feuille morte*).

It would be very useful to have a glass which transmitted all visible light, making it colourless, and absorbed all ultra-violet light. The only way of getting near this is to put very expensive rare-earth oxides (such as cerium oxide) in the glass, but even these give a pale yellow colour if sufficient is added to give a useful degree of ultra-violet absorption.

There has been very little published work on the effect of light on various products, perhaps because, as mentioned earlier, there is very little natural ultra-violet light about in the UK most of the time. (A possible source of some artificial ultra-violet light is the fluorescent lighting used in shops and window displays.) We do not need to consider light of wavelengths less than 280 nm, as even 'colourless' glass is opaque to such wavelengths; in any case, packers would have more serious things to consider if there were any such radiation about, as it could be sufficiently lethal to kill off the consumers before they managed to open their bottles!

The normal colour of amber glass used in the UK has a transmission at 500 nm of between 30% and 50% for a glass of thickness 2 mm. This is the transmission specified in BS 5144 for cider flagons and is also the same as that recommended by the Brewers' Society for beer bottles. The United States Pharmacopoeia specifies that a light-resistant container shall not transmit between 290 and 450 nm more than a specified percentage transmission, this transmission ranging from 10% for large containers to 50% for small ampoules.

Chemical properties

For the great majority of users, glass containers can safely be regarded as completely inert, that is as being completely unaffected by and having no effect on their surroundings. Indeed most packers and consumers have learned to look on glass as a material able to store any liquid or solid indefinitely without deterioration. The standard examples of this used to be the old bottles of Napoleon brandy and vintage port, but in modern times, there are many more mundane foods which thrifty housewives expect to stay in good condition in glass almost indefinitely.

In the last generation, the chemical resistance of bottle glass has been improved still further. When bottle-making machines were first introduced, the glass compositions were adjusted solely to give a glass which could be easily worked on the machines; such a glass might contain up to 20% of soda (Na_2O). As manufacturing techniques and knowledge of glass chemistry have improved, the compositions have been continually altered to give better chemical resistance, and it is only during the post-

F

war development period that the present composition, best both for efficient high-speed manufacture and good chemical properties, has become established. The Na_2O has been reduced to about 14%, and other ingredients such as alumina have been added to improve the resistance.

Strictly speaking, however, there is no such thing as absolute inertness, and all materials react with each other, but at vastly different reaction rates. The only liquid which reacts with glass rapidly at room temperature is hydrofluoric acid. Water and aqueous liquids have a reaction rate which is just measurable at room temperature, but for organic liquids (solvents, oils and other hydrocarbons) no reaction is detectable. The mechanism of the reaction with water is that a trace of the hydrogen in the water (H_2O) tends to diffuse into the glass, displacing an equivalent amount of sodium (Na), which is released into the water to form a solution of sodium hydroxide (NaOH). In other words the water will become slightly alkaline.

To give an idea of the rate of reaction, when storing aqueous liquids in bottles, in six months at $20°C$ the weight of sodium extracted from 100 cm² of glass surface might amount to 5 mg. Consider, for example, a cylindrical bottle containing 1 litre of water; its surface area in contact with the water would be about 500 cm², so the sodium extracted in six months would be 25 mg. The effect on 1 litre of water would be to increase its sodium hydroxide content by 0·005%. The reaction will go a little faster if some sodium hydroxide is present in the water to start with, but in this case the effect of gaining a little more from the glass is not likely to be significant.

The cases where the reaction between water and glass cannot be ignored are mainly those in which repeated high-temperature sterilisation of filled bottles (at 100 to $120°C$) is carried out. The times required to extract 5 mg of sodium from 100 cm² of glass at different temperatures are approximately as follows:

At $20°C$: 6 months
At $100°C$: 4 hours
At $120°C$: 1 hour

MEASUREMENT OF CHEMICAL RESISTANCE

There are basically two methods of estimating the rates of reaction between glass and water, using either whole containers or crushed glass ('grain tests'). In all tests the main idea is to exaggerate or accelerate the conditions so that some measurable effect can be obtained in a short time. In grain tests, the effective surface area is made very large by grinding, and this is a useful technique, for example when investigating different glass compositions. The user of bottles is more concerned with how the actual container surface will behave, and a test on a whole container filled with liquid is obviously more realistic.

Tests have to be carried out at high temperature; a steam autoclave can be used to give a temperature up to 125°C. Standard methods of test of this type are given in the United States Pharmacopoeia, the European Pharmacopoeia and in the British Pharmacopoeia.

However, since the chemical durability of glass depends directly on its composition, and as the composition is controlled very closely by the bottle maker, it is very unlikely that a glass packer will find any need to carry out tests of this sort. In this connection the EEC is developing a Directive controlling the chemical durability of all packaging materials which come into contact with food. This will undoubtedly include a test for glass containers, but the reason for this will be the need to be seen to be treating all packaging materials alike, rather than because of any genuine problem.

VISIBLE EFFECTS OF CHEMICAL REACTION

In the high-temperature treatment of bottles there may occasionally be a visible effect on the glass surface. After prolonged or repeated treatment in contact with water or alkaline solutions, a thin film may develop on the surface, consisting of what is left after some of the sodium (and a little of the other main constituents) has been replaced by hydrogen. This film is no longer a true glass; its physical properties are more like those of a plastic film, and being extremely thin (about 0·0005 mm) it may be visible when the bottle is empty because of the interference colours it produces, like the colours in a thin oil film or a soap bubble. Eventually small pieces of this film may become detached from the glass surface and appear in the liquid. These pieces of transparent film can catch the light and may be mistaken for small crystals or even pieces of glass; they are sometimes referred to, incorrectly, as 'glass flakes'. Some years ago, this phenomenon used also to occur occasionally during long-term storage of neutral or alkaline liquids at room temperature, but it has been virtually eliminated by the improvement in glass compositions.

If empty unsealed bottles are stored for long periods in a place where the humidity is high and the temperature varies, a slight chemical reaction may occur with the water which periodically condenses and evaporates on the glass surface. After some months the effect may become visible, because the reaction products accumulating on the surface will tend to make it lose its original brightness; this is the phenomenon known as weathering. The original condition can usually be restored by washing.

There are no reliable accelerated tests for investigating these visual surface effects, and experience must be the guide. (The bottle maker's experience is, of course, available to the packer.)

'SULPHATED' GLASS

If for any reason there is a risk that the chemical resistance of normal glass may not be sufficient, a special treatment, known as 'sulphating', may be given to the bottles during manufacture. This removes most of the sodium from the glass close to the inside surface, and so largely prevents further removal of sodium in service. While the bottle is at about 500°C the interior is filled with the acid gases sulphur dioxide and sulphur trioxide. (These can be produced by decomposing some ammonium sulphate inside the bottle, or by injecting the gases directly.) At this temperature the gases encourage very rapid diffusion of sodium out of the surface layer of the glass; the sodium is converted into sodium sulphate, and this remains on the surface as a 'bloom' which can be washed off with water.

Bottles treated in this way are widely used for such products as transfusion fluids, and for storing extremely alkali-sensitive drugs. The performance is improved about 100 times over that of untreated glass.

5. Choosing or designing a glass container

This chapter is written mainly for packers who have just decided to use glass containers for their products. Who do they do next?

The first question is whether to use one of the established 'standard' bottles, or to start from scratch with a new design. It is unusual for a new design to be necessary for solely technical reasons; it is very likely that there will be at least one existing design which will meet all the technical requirements. Bottles of established design possess the advantage of having already been proved in use, and methods for handling and packing them efficiently will have been established. Also if the design in question has been adopted as a 'general trade' or 'standard' bottle, it should be available for use at short notice, and if necessary in fairly small quantities.

At present, most packers have a fairly free choice of the unit quantities in which to sell their products, and therefore of the sizes of bottle to use. The 1963 Weights and Measures Act laid down mandatory size ranges only for milk, instant coffee, honey and jams. Pressure, however, from consumers and governments is steadily tending to change this, and it is likely that in the future many more products will have to be marketed in prescribed sizes. So this will be one thing less for the intending packer to worry about – but instead he will have to decide whether or not to achieve the required accuracy of fill by the methods applicable to the use of Measuring Containers. This point is discussed in detail later in this chapter; the main point here is that the trend towards prescribed sizes and prescribed filling methods is likely to accelerate moves towards design standardisation, and so the intending packer will be more likely than before to find that an exactly suitable container is available 'off the shelf'.

Established designs

The main present source of standards is of course the British Standards Institution. In the future their efforts may be supplemented by various international bodies, such as the Centre Technique Internationale d'Embouteillage, the Comité Européen de Normalisation, and the International Standards Organization.

In addition to these there are several 'industry' standards, usually arrived at by informal collaboration between groups of users assisted by their suppliers. Also in this category we may list some effectively standard designs which have evolved without any deliberate effort on anyone's part: we may call these 'traditional' standards.

Finally, some individual glass manufacturers have their own ranges of general trade bottles, and will maintain stocks of them for immediate delivery. In the past these have mainly been used for chemical and medical products, but as indicated earlier they are likely to increase in scope and importance, particularly during the transitional stages of metrication and size rationalisation. In many cases a glass maker's stock bottle may, if it does its job well, progress to becoming an industry standard and thence to a British Standard. The British Standard cider flagon (BS 5144) is a recent example of this process, and indeed it seems quite clear that this is the right and proper way to arrive at national standards.

BRITISH STANDARDS

Complete bottle specifications can be prepared and published by the British Standards Institution if a request is made by a number of BSI members. The committees who carry out this work do their best to consult all the organisations likely to be directly or indirectly involved with the bottle, including the glass manufacturers, packers and consumers, suppliers of closures and other packaging components and machinery, research organisations, government departments and overseas standardising bodies. BSI specifications usually cover all relevant dimensional and physical characteristics, with appropriate tolerances and means of verifying them.

So far the following British Standards have been published, and others are in the course of preparation.

BS 795   Ampoules (small containers made from glass tubing).
BS 830   Winchesters* (80 and 90 fl oz capacity).
BS 1679/2 Containers for pharmaceutical dispensing. This comprises ranges of oval and rectangular tablet bottles, also cylindrical jars with wide necks for powders and ointments.
BS 1679/6 Glass medicine bottles.
BS 1679/7 Ribbed oval glass bottles.
BS 1777   Honey jars ($\frac{1}{2}$ and 1 lb size; brimful capacities 163·5 and 327 ml).

* This name is used traditionally to describe a tall cylindrical bottle with a roughly hemispherical shoulder and a short narrow neck.

BS 2812/1   Milk bottles of the aluminium cap type.
BS 2812/2   Sterilised milk bottles. (The capacities of these milk bottles
             are dealt with in a separate Standard: BS 1925.)
BS 2463   Transfusion bottles for medical use.
BS 4590/1   Non-returnable soft drink bottles with crown finish (capa-
             city 0·33 litre or 12 fl oz).
BS 5144   Cider flagons (capacity 40 fl oz).

OTHER STANDARDS AND STOCK DESIGNS

It is of course not possible here to catalogue all general trade bottles which are available to any user, but they include the following main groups:

1. Traditional designs of spirit bottle. In the main these comprise tall round bottles for 750 ml and greater, flasks for some of the smaller sizes, and various designs of miniature. No formal attempt has yet been made to standardise any of these.

2. Traditional designs of wine bottle. Most of these are early candidates for adoption as British Standards, as soon as the metric capacity requirements are finalised. (The shapes and dimensions for hock, burgundy and claret bottles, etc., have already become effectively standardised; for fortified wines some alternative styles are available.)

3. The four sizes of returnable beer bottle used by all the British brewers. These have nominal capacities of about $\frac{1}{3}$, $\frac{1}{2}$, 1 and 2 pints. Standard designs were provisionally approved by the Brewers' Society in 1967, but they did not immediately progress into British Standards because of uncertainty at the time about Government metrication policies. The 0·33 litre bottle in BS 4590/1 is also suitable as a beer bottle, and similar styles of bottle with capacities of 0·25, 0·75 and 1 litre are also available (possibly not always in amber glass).

4. There has not been so much need for standard returnable bottles for carbonated soft drinks, as most bottlers prefer to use proprietary designs, but there are in effect standard designs in the usual sizes for 'mixer' drinks, and in the 0·75 litre size. Non-returnable bottles are available in the same sizes as for beer.

5. 1 lb jam jars, brimful capacity 370 ml. The original design of these was organised by the Food Manufacturers Federation before any light-weighting was carried out, but the current ones are still effectively standardised.

6. Winchesters, similar to those in BS 830, but with a wider range of capacities down to 50 ml.

7. Plain oval-section bottles for general purposes, and similar shapes with fluted panels for poisons.

Designing a new container

In spite of the standards available, the majority of glass containers used today are made to designs developed for an individual packer, product or market. This is not surprising, because as discussed in Chapter 1 the aid to product recognition or brand recognition provided by a distinctive package design can be a valuable marketing plus. So let us consider what the designer needs to know, how he produces his design, and how the design is turned into reality.

The designer, whether he is employed by the glass manufacturer or not, will welcome the opportunity of a design brief which tells him every-thing about the packaging requirements but which gives him a completely free hand to demonstrate his creativity with a striking new bottle shape. Alas for him, this does not happen very often because in most cases the new bottle is the heir or successor to a previous one, and needs to retain a degree of family likeness. And he has to take into account the dimensional and shape characteristics which will best suit the type of glass forming process to be used, otherwise the manufacturing costs will suffer. Never-theless, and in spite of the fact that many thousands of designs have been produced before, it is still possible for a good designer to produce a con-tainer which is visually fresh while being just as functional and economic as its predecessors.

The design work can therefore be safely entrusted to the glass manu-facturer's design department; alternatively the packer may employ an independent designer in the early stages. Both methods can and have been perfectly satisfactory, if carried out properly and with due regard to the technical requirements as well as to the appearance. The manufacturer should always welcome new ideas from independent designers, but there are many details of design which cannot be settled without the manu-facturer's specialised knowledge. The earliest possible co-operation between maker and designer is therefore highly desirable, and should not be discouraged for any fears that the designer will be 'talked out' of his original ideas.

INFORMATION REQUIRED

Before the designer starts work, he must establish as fully as possible which design requirements have already been fixed and for which aspects he can use his skill and experience to make proposals. The following check-list should cover all the relevant information, and each point is discussed in detail.

1. Bottle capacity.

2. Type of substance to be packed.

3. Details of temperature and pressure changes likely during processing or storage.

4. Type of closure.

5. Vacuity or headspace required. (If this has not been decided, it can be calculated from information under headings 2, 3 and 4.)

6. Colour of glass required.

7. Any requirements or limitations about shape, dimensional proportions, etc.

8. Plans for labelling or decoration.

9. Multi-trip or single-trip usage.

10. Any special user requirements (ease of holding, pouring, etc.).

11. Any legal restrictions or requirements, particularly those concerned with Weights and Measures, and the use of bottles as Measuring Containers.

I. BOTTLE CAPACITY

If the bottle is intended for a liquid, the capacity will usually be expressed as the volume which the bottle should hold when filled to a stated filling height (or occasionally when filled brimful). If the user's requirements are stated in terms of a certain *weight* of material, or a certain *number* of articles, this must obviously be converted into a volume for the purposes of design and specification.

The stated bottle capacity will be taken as the *mean* value required, and the permitted capacity tolerance will be applied equally above and below the mean.

Some liquids and semi-liquids are hot when filled, and contract as they cool; other liquids are filled at temperatures close to their freezing point and may expand. This will have to be taken into account when calculating the capacity required.

British glass manufacturers now work exclusively in metric units, so the capacities, if expressed by the user in other units, will be converted to millilitres or litres, depending on whether the capacity is below or above one litre. Conversion factors are given in Appendix A; care should be taken to distinguish between British and American fluid ounces, pints and quarts, and also perhaps between drachms and drams.

In some cases there will be a legal or commercial reason for embossing the capacity on the bottle itself, and this is best done in the same units as in the bottle specification itself, i.e. millilitres or litres. If this marking is to be the *sole* content indication on a bottle sold in UK, then from 1978 it will have to be stated in both imperial and metric units to comply with the Weights and Measures (Marking of Goods) Regulations, 1975. (The Regulations, however, are intended to apply primarily to printed statements on packages, including bottle labels.)

As we noted at the beginning of this chapter, there is an increasing number of British, EEC and foreign regulations controlling the weights or capacities in which products may be sold, and hence the capacities of the bottles used. A summary of current requirements is given in Appendices B and C.

Those interested in the history of packaging will be pleased to see that the traditional British 'reputed quart' of 26⅔ fl oz, equal to 0·75 litre and to the USA ⅘ quart, is still retaining its dominance in the wine and spirit field, although challenged by the 0·7 litre favoured in some other parts of Europe.

2. TYPE OF SUBSTANCE TO BE PACKED

Ideally the designer should know exactly what material is going into the bottle, and if possible he should have a sample of it, to assess its appearance in relation to that of the bottle. If it is a liquid, he should know enough about its properties to be able to estimate its thermal expansion and vapour pressure at any temperature. This information is required for calculation of vacuity (see item 5).

The chemical properties of the material may affect the choice of closure (see Chapter 9).

3. TEMPERATURES AND PRESSURES

The bottle will probably experience some sudden temperature changes, for example during washing or filling, or during subsequent heat treatments. The designer must ensure that the bottle has adequate resistance to withstand these thermal shocks. (This is not likely to cause any difficulty with modern bottle designs.)

Heat treatment after filling and sealing will cause an internal pressure rise, and the designer must know enough about the process to ensure that the bottle and the closure can resist the highest pressures produced. This applies particularly to beer, in which the carbon dioxide causes a pressure in the bottle of up to 6 bars during pasteurisation. On cooling, the pressure falls to about 1·5 bars. The sterilisation of uncarbonated liquids such as milk and transfusion fluids may produce pressures around 3 bars. Carbonated minerals are not usually heat-treated after filling; for these the highest pressures produced will be either during filling, when the pressures may temporarily rise to around 5 bars, or during storage in warm conditions.

During normal transport and storage in the United Kingdom, bottle temperature may occasionally rise to around 40°C. For bottles which are exported to hotter countries, it is reasonable to allow for a temperature rise to about 50°C or even higher in a few exceptional areas. Temperatures over 40°C are not likely to occur in this country, except occasionally for bottles displayed in sunny unventilated shop windows. The corresponding pressure rises may be insignificant or serious, depending on the vacuity and the properties of the liquid.

The figures quoted above are intended as examples only, and not as exact data to be used by the designer.

4. TYPE OF CLOSURE

The technical considerations which lead to a choice of closure are discussed in detail in Chapter 9, and of course the design aspects of the closure – appearance, colour and decoration – should ideally be considered as an integral part of the package design. As always, cost may be a significant factor in closure choice, also the availability of capping machinery.

Each type and size of closure has its own glass finish specification, most of which have been published by the Glass Manufacturers' Federation, or as British Standards,* or they can be obtained from individual glass or closure manufacturers. To fit a finish on to a bottle design the main consideration is that the diameters should match at the level where the finish mould (neck ring) meets the body mould. Also it is usually desirable to ensure that the diameter of the closure is less than that of the bottle body, otherwise the closure is liable to be damaged or even displaced during handling.

5. VACUITY

Bottles are rarely filled to the brim; it is generally necessary to leave a small space in the neck to allow for expansion. This is called the headspace, expansion space or air space, and when referred to quantitatively the term vacuity is used.† Vacuity is defined as the volume inside the sealed container which is not occupied by the product, and it is expressed as a percentage of the volume of the product. Thus if a bottle of brimful

---

* BS 1918 Part 1 Screw thread finishes
            Part 2 Crown finishes.
   † The term 'ullage' should not be used for glass containers. Its meaning in some trades when applied to barrels and drums may be quite different from that of vacuity.

capacity 110 ml is filled with 100 ml of liquid, and sealed with a cork which occupies 4 ml of space in the neck, the vacuity is 6%.

Vacuity is not usually critical when the contents are solid, but for most liquids it is an important factor of design. When any sealed container is heated, its internal pressure will rise, and the extent of the rise can be controlled by providing sufficient vacuity. The limiting factor for a particular pack is either the strength of the bottle or the leakage resistance of the closure, and both need to be known. Also the maximum temperature which the pack will reach during processing, or at any time subsequently, is relevant.

For carbonated beverages and most other aqueous liquids, including those containing alcohol, suitable vacuities are usually in the range 3 to 8%. (It is desirable for retail packs that the figure should not be needlessly large, or the bottle may not look full.) Organic liquids, especially the more volatile ones such as ether, may require 10% or more. Vacuum-sealed food packs need vacuities in the region of 6 to 12%, in order to maintain the effectiveness of the vacuum seal during heat treatment and afterwards.

The above figures are intended only as rough guides, and exact calculations should be made or advice sought from a closure expert whenever designing a pack for a new product. Methods of calculation and other data are given in Appendices E and F.

6. COLOUR OF GLASS REQUIRED

The colours available have been described in Chapter 2. Colourless glass is a natural choice if a good-looking product is to be displayed for sale, but coloured glass may be chosen for various reasons.

For some products a particular glass colour is 'traditional'; this applies particularly to the various classes of wine. There is no technical reason why, for example, hock should be bottled in amber glass and moselle in dark green; presumably these traditions became established in the early days of bottle making, when there was perhaps only one colour of glass made in any one area. Technical needs do arise if a product has to be protected from light of particular wavelengths, and the characteristics of suitable glass colours were given in Chapter 4.

7. SHAPE AND DIMENSIONS

It is rare for a designer to be given a completely free hand in choosing shape and dimensions: usually there are clear limits imposed by the types of machinery to be used or by other constraints.

If automatic machines are to be used for removing empty bottles from cases, or putting filled ones into cases, particular attention must be paid to the design of the neck and shoulder. Such machines usually hold the bottle with spring-loaded fingers or joints (or by pneumatically expanded grippers), and a projection or step must be provided at the right height on the bottle neck. (The bottle maker, too, normally handles the bottle in this way between the two stages of bottle forming, and it is sensible if the same design feature can serve the needs of maker and packer.)

Bottles usually travel in an upright position on filling lines, the motion being controlled partly by the moving conveyor belt and partly by the forces applied by adjacent bottles on the line. Cylindrical bodies, or bodies with vertical flat sides, are best for this purpose, but a waisted or recessed body is also satisfactory provided that there are two different heights on the body at which the diameter or width is equal. This point is illustrated in Fig. 5.1. Example (f) is bad because the upper point of contact occurs at a sharp edge, where the strength will be relatively low.

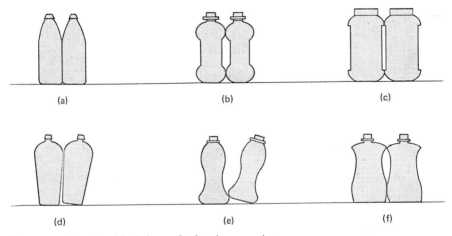

Figure 5.1.   Good and bad shapes for bottle conveying

Exactly the same design considerations apply to bottles which are to be bulk-palletised.

The stability of the individual bottles is also important. Any tendency to tip over must be avoided when the bottle is made to travel round bends, or when a conveyor belt suddenly starts or stops. The controlling factor is the height of the centre of gravity in relation to the width of the base. The force required to upset the bottle is proportional to the tangent of the angle θ shown in Fig. 5.2, so for greatest stability θ should be as large as possible. For practical purposes it is usually sufficient to say that the ratio of base dimension to overall height should be as large as possible.

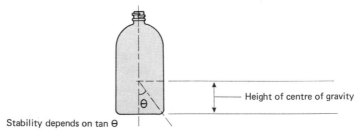

Figure 5.2.   Bottle stability

Resistance to upsetting by a sideways impact near the top of the bottle also depends on the shape in much the same way, but in this case the weight of the bottle, and contents if any, helps stability. This is the reason why a bottle sometimes tends to become less stable if its weight is reduced, unless its shape is also altered to increase the angle θ.

Bottles of low stability can of course be handled satisfactorily, but usually at the cost of more complicated handling machinery and lower speeds and efficiencies. Bottles with oval or flat cross-sections are more stable in one direction than another, and they may be handled quite efficiently if guides are placed on the filling line to stop them from tilting in their less stable direction.

If the filled bottles are to be displayed for sale without outer coverings, the retailer may want to stack one on top of another. With screw-capped jars this can be arranged very simply, for example, by an arrangement such as shown in Fig. 5.3.

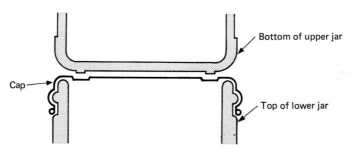

Figure 5.3.   Stacking method for screw-capped jars

8. LABELLING AND DECORATION

If paper labels are to be applied, the designer needs to know their intended size, shape and position in order to ensure that the bottle surface will be suitable. The best labelling surface is a cylindrical one with a large radius, so that the label can be smoothed round the curve with a simple wiping action. Flat panels are also suitable, but a small amount of convex curvature is helpful even on a nominally flat surface, in order to assist the stretching of the label and to avoid difficulty if the flat panel sinks slightly when the bottle is blown. It is obviously not possible to apply labels to a spherical surface, or to any convex surface which curves in two directions; a concave surface is also out of the question, at least for automatic labelling.

A slightly recessed labelling area on a cylindrical body (see Fig. 5.1.c) can be a very useful way of protecting the label in use.

Much the same design considerations apply to bottles which are to have a permanent vitreous enamel label applied by silk screen, or any form of gold or other metallic decoration, or a design applied by transfer. The printing of unfired colours on glass is also possible, and if this is done by rollers, a cylindrical bottle is essential.

9. MULTI- OR SINGLE-TRIP USAGE

For many types of bottle the question of trippage does not arise, because there is some overriding reason why the bottle will be used by the packer only once. This applies, for example, to all exported goods, many pharmaceutical and toilet products, and for any product where no system exists for collecting the empties.

For some products, notably beer, cider, minerals and some dairy produce, the packer or designer may wish to choose between two clear alternatives. One is to use a returnable bottle: for this the requirements are that the bottle should be capable of standing up to repeated rough use with the minimum of damage, and of maintaining its strength at a safe level indefinitely. The economic justification for this approach is that the container cost per trip is so small. Some of the bottles will, of course, be expected to perform many more than the *average* number of trips, and the strength of the bottle must be sufficient to make this possible. If the bottles have to withstand internal pressure, it is usually necessary to allow a factor of about three between the maximum pressure expected in service and the minimum strength of the new bottle.

The other alternative is to use a bottle for which the weight and strength is reduced to the minimum needed for one safe trip. The strength factor for one-trip pressure bottles can be reduced to about two, compared with three for returnable bottles.

Both types of bottle have their place in glass packaging, and the packer will not usually find any difficulty in deciding which best suits his particular purpose. Much depends on container cost, number of trips possible, transport distances, efficiency of collecting empties, together with the relative handling costs on the filling line. But the final word may lie with the retailer or consumer.

It must be stressed that the choice we have been discussing is between a bottle which will make one trip and a bottle which can make as many as possible. It is understandable for the packer to wonder whether a bottle of intermediate strength and cost might perform sufficient trips to make it more economical than either extreme; this is not impossible in theory, but in practice it rarely works out this way. The most valid reason for considering an intermediate is if existing multi-trip bottles are so strong that virtually none is ever broken, or if a single-trip version is so weak that significant breakage occurs during the one trip. This, however, is a matter of correcting a bad or out-of-date bottle design rather than altering the intended function of the bottle.

10. USER REQUIREMENTS

So far we have considered the information needed by the designer to ensure that the bottle can be packed efficiently. In addition, the designer must be alive to the requirements of the ultimate user, so he needs to know what is likely to happen to the bottle after purchase. For example, will the bottle be handled and opened with wet hands? What are the pouring characteristics of the product, and how much will be used at a time?

Attention to these matters is important; at the same time the designer may need to be a little wary in his recommendations, because the public do not invariably choose, or like to pay for, what has been found to be technically the best for them.

11. LEGAL REQUIREMENTS

In a sense all the foregoing sections convey a legal implication, as the glass maker and the packer share a responsibility that the goods offered for sale to the public should be 'fit for the purpose required', otherwise they may be in breach of the Sale of Goods Act, 1893. Also the Health and Safety at Work Act, 1974, requires that the glass maker shall ensure, so far as is reasonably practicable, that the bottles he supplies shall be safe when properly used, and that the packer understands how to use them properly. And finally, any carelessness in the design, manufacture or usage of the bottles which leads to loss or injury may give rise to an action for Negligence.* Apart from these general obligations, there are several specific requirements:

1. *Control of bottle contents.* Under the 1963 Weights and Measures Act, it was primarily the packer's responsibility to ensure that the bottle he sold contained not less than the amount he declared on the label (for those products subject to the Act). Only if the bottle departed so far from capacity specification that the packer's task was impossible might the bottle manufacturer become involved in a prosecution. But this situation is changing, because the 1963 Act has to be significantly amended not later than 1979, to bring it into line with two packaging Directives accepted by EEC member states in 1974. These Directives aim primarily to regulate the contents of packages sold in one EEC country but packed in another country; it is very likely, however, that the amended British legislation will apply generally to goods sold in this country including home-packed ones. The basis of the EEC Directives is briefly as follows:

(a) Packers of most liquids for drinking ('liquid foods') are assured that if they choose to pack their product in one of the standard sizes, and control the contents within stated tolerances, then their pack cannot be refused right of sale in any EEC country. (Strictly speaking, the choice of sizes is optional, not mandatory, but clearly no sensible export packer would opt for a non-standard size.)

(b) The content to be stated on the package is the average, not the minimum.

(c) The main point at which legal control is exerted is at the time of packaging, not the time of sale. Government Inspectors will visit factories to make tests, and in some circumstances may examine inspection records.

(d) The packer should control his contents *either* by using a legally approved volumetric filling machine,† *or* by carrying out regular statistical checks on the volumes he is filling, *or* by using Measuring Containers, and simply controlling the filling height or the volume of the headspace in the filled bottles.

The corresponding UK legislation is likely to follow this closely, probably with the extension that sizes previously permitted or required in the

---

*Also a draft EEC Directive is under discussion, which seeks to impose on manufacturers financial responsibility for losses, regardless of any culpability.

† This applies to products sold by volume (e.g. liquid food). The Directives which relate to foodstuffs sold by weight will require the use of a legally approved weighing machine or regular statistical checking of the weights filled.

G

UK may continue to be used in addition to those in the EEC lists. Also the inspection will be carried out by the existing Local Authority Inspectors.

The legal requirements of the alternative content control systems are set out in Appendix C. Most liquid filling machines are designed to fill bottles to within a fixed distance from the top, so the most prevalent measuring container is likely to be of the type which specifies its mean capacity at a stated filling height. Glass manufacturers should be able to comply with the capacity tolerances, and so agree to apply the measuring container symbol, with most straightforward designs having a cylindrical or similar cross-section. Undoubtedly, however, it will be found that some of the more complex existing designs, especially in some of the very large sizes, are too variable in capacity to meet these requirements; in these cases the packer will still have the option of the two alternative control methods, or if the measuring container system is particularly important to him he could consider changing to a more regular design of bottle.

2. *Packaging of poisons.*   Under the Poisons Rules (No. 25) made under the Pharmacy and Poisons Act, 1933, a container in which poison is supplied must be impervious to the poison and sufficiently stout to prevent leakage arising from the ordinary risks of handling and transport; and if the container is a bottle (with some exceptions) the outer surface must be fluted vertically with ribs or grooves recognisable by touch.

3. *Packaging for export.*   Bottles which are to be sold in another country must obviously comply with any relevant laws or regulations of that country. For example all spirit bottles imported into the United States must be embossed with the words 'Liquor Bottle' and the country of origin. There are many local requirements in other countries covering such things as the use of pilferproof seals, permitted headspace and strength characteristics.

4. *Patents and Registered Designs.*   The designer should make sure that the new container and its closure does not infringe any registered patent or design. If he thinks that a feature of his design might be protected by prior registration, he can have a search carried out at the Patent Office in London. If the container is for export, then a search would be necessary in the country where it is to be sold. Infringement of a registered patent or design might lead the manufacturer into a civil action by the owner of the infringed registration, with the danger of a court injunction restraining further manufacture, or the liability to pay a sum of money as damages, or both. Even where there is no registration, care must be taken that the

new container does not resemble another one so closely that it could lead to confusion in the mind of a potential purchaser. This is only likely where the contents also are similar; the packer might then be sued for 'passing off' his product as that of another.

Manufacturing considerations

The discussion so far has been concerned with planning a bottle which will do its job efficiently; it is nearly always just as important that it be designed so that it can be manufactured efficiently and economically. In practice there have been very few cases indeed where it has been found commercially worth while for the bottle maker to be given the problem, and the packer to bear the increased cost, of making a really difficult bottle. Difficulty in this sense relates primarily to the rate at which the bottles can be produced, the amount of glass required and the proportion of finished bottles which have to be rejected as defective. The cost of making and maintaining complex or irregular mould cavities can also be very significant.

So let us consider what the glass maker has to bear in mind when he turns the initial design proposal into a final manufacturing specification.

SHAPE AND GLASS DISTRIBUTION

We have referred before to the importance of glass distribution in ensuring that a bottle has adequate strength in all conditions. With the aid of modern computer techniques one can get quite close towards working out the optimum glass distribution for a particular design of bottle to operate efficiently in specified conditions. In practice however, although the glass maker may control glass temperature and machine timings, and may choose the best possible gob shape and parison design, he still only has a limited degree of control over how the glass moves inside the mould, and he is unlikely to achieve perfect glass distribution consistently, if at all. His primary concern during manufacture will be to try and prevent sources of weakness such as local thin spots and local abrupt changes in glass thickness.

The glass gob is extruded through a round orifice, and so is roughly cylindrical. The parison usually retains a round cross-section, but probably increases in diameter steadily from one end to the other (Fig. 5.4). So obviously the bottle shape with the best chance of symmetrical glass distribution will also be cylindrical or nearly so. If we consider such a bottle being blown in its mould, and then imagine suddenly replacing the

Figure 5.4. Typical parisons and resultant glass distributions

cylindrical mould by another mould of different shape, it is easy to visualise how the glass would have to be stretched to reach the further parts of the mould. This is shown in Fig. 5.5, which also illustrates the effect on glass distribution of abrupt changes in external section. (With bottles of markedly non-round cross-section the parison may also be made non-round to reduce the effect described, but its ability to compensate is rather limited, because it tends to resume a cylindrical shape as soon as the final blowing starts.)

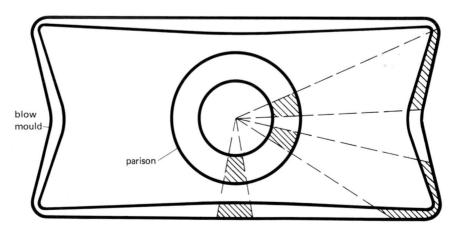

Figure 5.5. Effect of shaping a cylindrical parison into a non-cylindrical bottle

So if bottle designs depart significantly from the round, they will require extra glass to compensate, and if the ratio of their major to minor axis exceeds about 4:1 they will be difficult to control at any weight.

Another design ratio to bear in mind is that of height to width. If this is too great it is impossible to maintain a symmetrical temperature distribution in the parison, which would cause further unevenness in distribution. For narrow-neck bottles optimum ratios are usually around 2·5:1 and if possible they should not exceed 4:1. For wide-neck jars made by the press-and-blow process, the optimum is about 2:1 and anything over about 3:1 will affect strength or productivity.

EFFECT OF SHAPE ON STRENGTH

We discussed in Chapter 4 the effect of cross-sectional design on internal pressure strength and the effect of profile shape on vertical load strength. The first of these is so crucial that it is virtually obligatory that bottles for pressurised liquids should have round cross-sections and should be essentially cylindrical. Also in any case where the design brief is to produce the most economical container possible there will be a strong incentive to stay with or close to the basic cylindrical shape, in order to get maximum strength with minimum weight of glass.

FINISH DESIGN

It should be remembered that the first action in bottle making is to form the finish by pushing the glass at the bottom of the gob into the neck ring; the time required to achieve this will be affected by the intricacy and depth of the finish design. Then the parison body is formed, and as soon as the finish has cooled to a rigid state the parison can be transferred

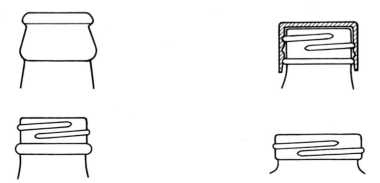

Figure 5.6. Types of supporting beads or undercuts

to the blow mould. Transfer may have to be delayed by slowing down the machine if the finish is too heavy in proportion to the rest of the parison.

At the bottom of the finish a suitable supporting bead or undercut ledge has to be provided, by which the parison is supported in the blow mould and the completed bottle lifted out of the forming machine. The same bead may be made suitable to assist mechanical handling of the bottle on the filling line (Fig. 5.6).

Thus glass manufacturers will be assisted if the finish is simple, streamlined and not too heavy or too tall. These points have been taken care of as far as possible in most standard finish designs, and in some cases alternative designs have been established to suit different styles or weights of bottle.

EDGES AND CORNERS

With some decorative designs there will be a strong urge to provide sharp angular edges and corners, in contrast to the usual flowing curves or perhaps as a derivative of the crystal cut-glass style. This may cause problems with mould manufacture and maintenance, and sharp angles of 90° or less are liable to cause a glass distribution problem as well as being prone to damage in use. Nevertheless there are often ways of achieving a desired visual effect; Fig. 5.7 shows that if the angle is made greater than 90°, or if the angle is turned in two steps to make a 'blake' edge, then quite sharp edges can be attempted. Of course the glass almost never produces an optically perfect angle, because the surface 'freezes' very rapidly as soon as it is cooled by the mould iron, and some radius of curvature is unavoidable.

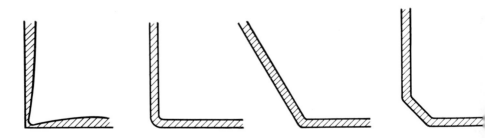

Figure 5.7.   Corner designs

SURFACE DECORATION

A great variety of surface effects are possible, and provided they are designed skilfully they should not cause any manufacturing difficulty or weakening. Indeed some forms of surface knurling are actually aids to strength maintenance, especially for lightweight carbonated beverage bottles which might otherwise be weakened by handling abrasion. (See Chapter 4 – *Resistance to abrasion.*)

Care must be taken when applying raised lettering or decoration close to the side seams of the bottle, as if too pronounced it may interfere with the opening of the two mould halves.

SHAPE CONTROL

A simple streamlined design of bottle which blows easily into its cylindrical mould should retain the mould shape accurately. By contrast, some complex designs fail to do this reliably because some of the air between the glass and the mould is sometimes unable to escape through the mould joints, but becomes trapped in recesses or corners. A cure for this which is partially effective is to drill tiny vent holes in the affected parts of the moulds.

Bottles with flat side panels have a different problem: they may be of perfect shape when first formed, but they tend to deform slightly (to sink or to bulge) as the bottle cools during annealing. This can be unsightly or interfere with labelling, and perhaps more importantly can upset the bottle's capacity control. It is often possible to restrict this effect by designing the bottles so that the panels although apparently flat are actually slightly bowed outwards.

THE FINAL DESIGN

It might be thought from the foregoing remarks that if the glass manufacturers had their way every bottle would be designed with the same old cylindrical shape and the same old proportions. But in fact their designers take a pride in exploiting the versatility of their medium as far as economics permit, and this will be confirmed by a glance at any manufacturer's showcase or bottle library. It is possible to keep fairly close to round cross-sections and still have a great deal of variation of shape and style, as shown by the examples in Fig. 5.8, and the many forms of surface decoration possible add further variety.

Figure 5.8. Distinctive designs with good cross-sections

Design procedure

The glass designer has not only to produce ideas but also to communicate them effectively to the prospective user, and to the prospective manufacturer. This is usually best done in a number of stages, each more detailed and precise than the last.

If the requirement is for a completely new design, without any constraints about previous packs or current house styles, it is probably best to start with simple sketches showing approximate plan and elevation or three-dimensional appearance. When provisional agreement has been reached, the manufacturer will estimate the volume of glass required to meet the strength requirements etc., hence the total external volume of the bottle can be calculated, and external dimensions estimated which will provide this volume. If any dimensional requirements or restrictions will arise in the usage of the bottle, they must obviously be taken into account at this stage.

It may now be helpful for the user to see a picture of what the final bottle should look like when filled, capped and labelled, and this can be supplied, either in black and white or coloured, as required. Alternatively, a three-dimensional model may be needed, and this can be cut from solid clear perspex. If such a model is dyed to reproduce the effect of coloured glass and/or coloured contents, then labelled and capped, a very realistic facsimile of the real thing can be obtained.

Next, a detailed specification drawing is prepared, showing all the dimensions, tolerances and finish details, and all the decorations or markings which will have to be reproduced in the mould cavity.

Finally, the parison mould and blow mould are designed and mould equipment made for a single head of a forming machine. In determining the dimensions of the blow mould, allowance has to be made for the shrinkage of the glass after the hot bottle leaves the mould. The initial moulds are then run under production conditions and some 'sample' bottles are produced. These bottles should be examined carefully by both user and manufacturer, because this is their last chance of checking that they are correct and satisfactory, before the rest of the mould equipment is made.

DESIGN CALCULATIONS

The essence of the designer's mathematical work is the calculation of a volume defined by various linear dimensions. This used to be a very lengthy task for anything but the most regular shapes, and for some of the most complex designs it was actually mathematically impossible to calculate the volume exactly. This led to the first use of bottle models, as their volume can be measured physically by 'displacement', i.e. by weighing in air and in water, following Archimedes' Principle. Indeed this method is still used sometimes to provide a check on calculations.

Modern electronic calculators have considerably shortened the time required for such calculations, often from several hours to a matter of minutes. It is now possible for design offices to be linked directly to full-size computers, so that not only can calculations be completed in seconds, but the computer can be programmed to adjust some or all dimensions until exactly the right volume is obtained.

In addition, it is quite possible for a computer to do the actual drawing work, either producing an outline on paper, or even projecting a three-dimensional impression on to a screen. It is even possible to instruct the computer to move the proposed bottle around on the screen, so that it can be examined from any angle. It seems rather unlikely, however, that this step will become regular practice in the real world of packaging, at least for some time to come. The main reason is that all the people who have to be involved in commercial decisions about new packages are usually in different parts of the country or of the world, and the quickest – and cheapest – way of communicating with them all still seems to be to post them a drawing!

More significant computer developments are taking place in the field of mould design and manufacture. Computers are being programmed to play a large part in the design of moulds and plungers. Then if other engineering information is provided, the computer will convert its knowledge into a punched tape which can be used to instruct and control the lathes employed for machining out the required mould cavity. This system of 'numerical control' in mould making should lead to greater accuracy and quicker delivery of sample moulds than has been possible before.

6. Planning and operation of bottling plants

A decade or two ago many people thought that the general state of development of UK packaging in glass lagged somewhat behind that of other countries, particularly the USA. There was probably some truth in this, but the main feature which tended to distinguish us from the others was the great diversity of products required for the British home market and the large variety of packs needed for worldwide export markets. As a result the usual need was for a very flexible filling line capable of doing several different jobs in succession, and there were not many packers in the happy position of being able to run a line continuously on one product in one pack.

In recent years however rationalisation has been the keyword, applied to products, packs and packaging companies, and there is now much more scope for the use of fully automated high-output lines. British packers have been quick to exploit their opportunities.

The traditional filling line of the past consisted of a washer, a filler, a capper and a labeller, possibly purchased from various suppliers at different times, and connected if space permitted with a straight-line conveyor belt. Many of these are still in efficient operation, and they can provide good flexibility if needed, but probably the more typical modern installation is a complex integrated engineering installation, of capital cost over £1 million, and apparently contravening many of the planning principles which the author enunciated in 1963.

Let it be said immediately that there was nothing wrong with those basic principles – of keeping the conveyor lines as level, short and straight as possible – and they should still be adhered to as far as possible. Fig. 6.1 shows a good example of this – a tablet filling line at The Boots Co. Ltd. But the requirements of some packaging processes now make changes of height and direction inevitable, and means have had to be found of doing this without damaging the glass or losing control of the movements.

The greatest changes have been caused because the methods of delivering the empty glass to the line and of repacking the filled glass have diverged considerably. If bottles are to be unloaded mechanically from

Figure 6.1.   Tablet filling at Boots, with a King 6-head counter/filler and cotton wool dispenser and a 4-head Dico capper

bulk pallet loads (see Chapter 7) the lowest possible unloading height from the floor is equal to the height of the complete load plus the height of the pallet-lifting mechanism, perhaps 3 metres or more. These then have to be brought down to the normal operating height of about 1 metre above floor level, and if the bottle is at all unstable considerable care is needed in designing the conveying system. Fig. 6.2 shows how this has been successfully achieved by Distillers Co. Bottling Services Ltd at Leven. The Stork depalletiser and lowcrator bring the bottles down along seven lengths of gently sloping conveyor without permitting any slipping or build-up of pressure.

The size and complexity of high-speed machinery now makes it increasingly difficult to link all the units in a straight line, and it is of course impossible when output is boosted by having twin fillers or labellers, etc., working in parallel. Fig. 6.3 shows how this has been dealt with at Quinneys Dairy Ltd, Redditch, where two Vickers-Dawson 48-head fillers are supplied from a single washer (on the right) and the filled bottles go to a single 3-head crater (in the centre). So means have to be found of changing the line direction without causing pressure or speed changes.

Figure 6.2. Johnnie Walker whisky bottles descending from Stork depalletiser at 300/minute

Figure 6.3. Dual line at Quinneys, with each Vickers–Dawson KA filler/capper running at 400 pints/minute

The fact that there have to be exceptions does not mean that the rules themselves can be forgotten, and it is still worth while summarising the main planning and operating principles which should be followed as far as possible.

Planning principles

1. *Design the line to suit the packs, and not vice versa.*   This is easy to observe when a completely new line is being planned, but there is often a temptation to modify a pack in some way in order to be able to use some existing equipment which is not quite suitable for it. This nearly always turns out to be a false economy.

2. *Evaluate the likely range of packs to be handled on the line, and plan ahead for efficient changeovers.*   The extent and likely frequency of pack changes should be considered carefully, as this may affect the choice of machinery. Again it is a question of choosing 'courses for horses'. The more sophisticated a filling line is, the longer it is likely to take to convert to a different pack, and time means money. In extreme cases it has sometimes been found worth while to have alternative machines in parallel or in tandem, so that the changeover can be instantaneous.

In addition to changes of unit pack there may well be variations needed in outer packaging. Also the line planner should consider as far as he can the possibility of future changes because of technological advances.

3. *Plan for optimum visibility and accessibility of the operating machinery.*   This helps effective operation and control, but it is also important for efficient maintenance and cleaning as well as for emergency adjustments and repairs. A straight-line layout is usually the best if space permits, such as shown in Fig. 6.4, but it is sometimes advantageous to arrange individual machines on spurs off the main line, particularly if they are being duplicated. Fig. 6.5 shows three Krones Solomatic labellers so arranged, with parallel infeed and return feed to the main line, in a way which gives unusually good access to the labelling machinery. This installation is at the Eichbaum Brauerei, Mannheim.

4. *Plan for maximum operator efficiency.*   Even though the number of operators required has been drastically reduced with modern automatic lines, it is just as important as ever that they should be able to carry out their jobs effectively, comfortably and safely. This means, among other things, giving proper attention to lighting, seating, space heating and relief arrangements, and providing appropriate guards, safety devices and

Figure 6.4. 300/minute whisky line at DCBS, Leven, with equipment by Hill, Stork, Metal Closures, Krones and Kettner

protective clothing, remembering especially protection of eyes and ears. In addition the operators should be thoroughly trained, and instructed exactly what to do – and what *not* to do – in the event of breakdowns, and they should have effective and responsible supervision.

The Health and Safety at Work Act, 1974, provides legal obligations which fit in very closely with these requirements.

Figure 6.5. Three Krones Solomatic labellers on a single high-speed line

5. *Plan for co-ordinated supplies of all the correct packaging components – and of the product itself.*   Nothing is more frustrating than having to stop the line because the labels have run out, or to find out the hard way the caps do not fit the bottles. There should where possible be stand-by supplies to switch to immediately if a batch of components seems to be giving trouble. It might be thought unnecessary to remind packers about continuity in the supply of product, but experience has sometimes shown otherwise. In particular, the arrangements for changing from one product to another, or from one supply tank to another, are sometimes responsible for quite long delays.

6. *Where necessary, plan for the reception of visitors.*   It should not be over-looked that a modern efficient packaging operation can be very impressive and fascinating for the potential consumers, and it may be desirable to think in advance about what they should usefully and safely see.

Operating principles

The operation of individual packaging operations is discussed in the following chapters; this section is mainly concerned with the line itself, that is the system for conveying the bottles between the several stations.

There are a few processes where the bottles are carried throughout in individual pockets or compartments, but the more usual and simpler method is to use a flat-top slat-chain conveyor. This has to be sufficiently smooth and stable in motion to allow the bottles to stand freely and steadily in any position relative to the slats, and the friction between glass and conveyor has to be sufficiently low to allow the bottles to slide easily when stopped or diverted from the conveyor's direction of movement. The chain must also be easily capable of positive drive, and resistant to wear and abrasion.

Stainless steel or other types of hardened steel are commonly used for conveyor slats, and their surfaces are often lubricated with soaps, detergents or specially compounded 'invisible' lubricants* to reduce friction. Alternatively thermoplastic chains are used, and if suitably made may not need lubrication. They are of course not as wear-resistant as steel.

Let us now consider some of the important operating principles for conveyor systems.

1. *Keep positive control of the bottles when cornering is necessary.*   The old method of using a fixed obstacle to deflect bottles on to a stationary 'dead-plate', and then pushing them on to another conveyor with the pressure

_____
* For example, 'Dicolube SL', supplied by Diversey (UK) Ltd.

of bottles queuing behind, is definitely not recommended for efficient and safe high-speed operation. The two basic good methods are shown in Figs. 6.6 and 6.7. In the first, jars of Robertson's marmalade are being deflected, by curved plastic guides, from a straight conveyor on to a rotating transfer disc. This is motor-driven to carry the jars along a circular path until they are deflected on to the right-angle conveyor. In the second example John Walker whisky bottles are being conveyed round a 90° corner with a side-flexing chain conveyor. Both methods work very well, but the latter is preferable for tall or unstable bottles because it is smoother and

Figure 6.6. Cornering at Robertsons, Bristol, with powered transfer discs

more positive, and does not rely solely on the guide rail to start the change of direction.

Finally, most packaging machines carry the bottles around a circular path, and it is sometimes possible to take the outfeed in a different direction from the infeed, and so use the machine to turn a corner. It is more usual however for the rotating table of the machine to be arranged tangentially to a straight-line conveyor, which means that the bottles have to turn through three part-circles before continuing in the same direction.

2. *Keep positive control when height changes are essential; don't use gravity.* Gradients of a few degrees can usually be accepted on conventional conveyors, unless the bottle is unusually unstable. But if a bottle has sufficient

H

Figure 6.7.   Cornering at DCBS, Leven, with side-flexing conveyor

space and gravitational force to start sliding on a sloping conveyor, it will continue to slide until it meets an obstacle, probably the next bottle. This too may start sliding and the whole system can get out of control. So gradients must be severely restricted, and also conveyor surface lubrication avoided in sloping sections.

Major height changes usually require conveyors with individual bottle pockets, but Fig. 6.2 showed a good example of what is possible with conventional conveyors when space permits.

3. *Keep bottle speeds as uniform as possible, and use controlled acceleration and deceleration when necessary.*   If the bottles can move at a uniform speed from one end of the line to the other without interruption there can obviously be no fallen bottles and no glass-to-glass impacts. This used to be fundamentally impossible, because the bottles were expected to join queues in front of each machine, and would then be accelerated to about double the general line speed while they went round the filler or capper etc. With modern high-speed machines it is becoming more nearly attainable, because the methods of infeed to the machines are now more positive, and every effort has been made to get the filling or capping heads close together (see for example Fig. 6.3) thereby bringing the circumferential speed nearer to that of the conveyors.

Some degree of acceleration will usually be needed in order to place the bottles stably on the machine pedestal, and the two basic methods are shown diagrammatically in Fig. 6.8. In the first the bottles are marshalled by a timing screw of constant pitch, then speeded up by the intermediate guide wheel which runs in synchronism with both the timing screw and the pedestals. The second and better method achieves the speed increase with an accelerating screw, then the bottles continue at constant speed around the guide wheel.

Feed screws are usually made from low-friction plastic materials, and they should be constructed accurately to match the bottles being handled.

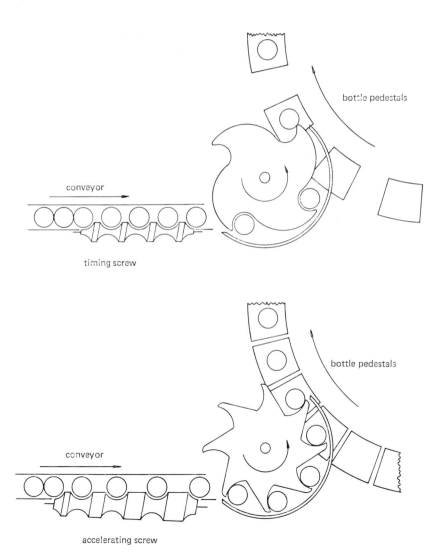

Figure 6.8.   Two types of screw feed

They can be used with advantage at any points on the line where exact positioning of the bottles is important; Fig. 6.9 shows a twin-screw arrangements taking J & B whisky bottles into the top-strap labeller at Strathleven Bonded Warehouses Ltd, Dumbarton.

Figure 6.9. Screw-feeding J & B whisky bottles into top-strap labeller at Strathleven, Dumbarton

4. *Plan to minimise the effects of machine stoppages.* If on a simple line one machine stops temporarily, the preceding conveyor will rapidly fill up with stationary bottles, possibly grinding against each other, then the preceding machines on the line will have to be stopped in turn. One way of avoiding this is to have a fully synchronised line driven and controlled from a single point. This is particularly suitable for bottles of awkward shape which might be difficult to handle otherwise, but it means that there is no way of catching up for lost time.

It is generally found better, as far as productivity is concerned, to make the line as flexible as possible by including accumulation areas at strategic points, where bottles can wait temporarily until the stopped machine restarts. It has always been useful to provide small tables on to which operators can lift out some bottles during machine adjustment, and this can easily be mechanised, as seen for example in Fig. 6.10, taken at Parke, Davis and Co., Inc., Pontypool. Normally the bottles travel in single file round the edge of the rotating table, but if there is a hold-up at the exit they go round again, and if necessary the whole table can be filled up. A moving belt is necessary in the centre of the table to guide the waiting bottles towards the exit when possible.

Figure 6.10.   Rotary accumulator at Parke Davis, Pontypool

On high-speed lines the accumulators may usefully be made much larger than is possible with a rotating table. Fig. 6.11 shows a straight-line system of very large capacity at Robertsons Ltd, Bristol. Normally the filled jars emerge from the cooling tunnel on the left, and travel to the right on the nearer part of the transverse belt. If there is a hold-up at the next station some of the jars are carried back on the farther side of the transverse belt and led on to the moving accumulator belt in the centre.

Figure 6.11.   Linear accumulator at Robertsons, Bristol

When the stoppage is cleared the belt changes direction and feeds the jars back on to the line.

On an ordinary non-synchronised filling line it is usually best to have each successive machine running about 5% faster than the previous ones, but if accumulators are provided the speed increases can usefully be greater than this, so that waiting bottles can be cleared more quickly. It is also important to adjust the speeds of the linking conveyors, and the belts should not be run any faster than is needed to maintain production.

5. Minimise friction and glass impacts wherever possible.   It might be thought that it does not do any harm to bang bottles about on the line, provided that they do not actually break. As explained earlier, this is absolutely wrong, because any visible or invisible damage to the glass surface will weaken it slightly, and may increase the risk of breakage later. And in any case it is a direct waste of energy. So attention to several details is important and worth while.

Bottles need sufficient space to travel freely along conveyors, but too much may allow spinning or jamming. Conveyors should therefore be provided with adjustable guide rails which lock firmly in position after adjustment. Metal guide rails of round cross-section are usually better than rails of flat section. Best of all are non-metallic or plastic-covered rails – especially on high-speed lines. Use of non-metallic or plastic-coated materials is also advisable for all other bottle-handling parts, such as star-wheels and screw feeds. Teflon (polytetrafluorethylene) is particularly suitable, because of its very low friction.

If it is not possible to construct the guide rails from non-metallic materials throughout, they should be covered with plastic at all bends and crossovers, and other places where bottles ride against them.

Particular care is needed when bottles travelling several abreast are being divided into parallel lines, or accelerated into single line. Stationary line dividers should be well cushioned, and the relative speeds of the separate sections of multiple belts adjusted carefully. Finally, if in spite of all efforts bottles occasionally fall over, arrangements should be made to eject them from the line.

The surface treatments applied by the bottle manufacturers help considerably in reducing line friction and abrasion. In the past this has only been true of single-service bottles, because the treatments did not survive multiple washing. So for returnable bottles it has sometimes been found useful to spray a lubricant, detergent or even plain water on to the empty bottles as they are fed into the refilling process. Now however surface treatments are being developed which will remain effective in multi-trip use.

Checking the characteristics of packaging materials

As referred to in Chapter 3, the manufacturer of the glass and other materials will have carried out regular checks on dimensions and other characteristics as part of his normal quality control process. So it should not often be necessary for the packer to make repeat measurements when the supplies reach him. If such measurements are required, however, either as a simple check against specification or as part of an investigation into a packaging problem, then it is helpful for the packer and the supplier to use the same methods of measurement, in case data need to be compared.

Glass container dimensions and their tolerances have been defined in ways which relate as directly as possible to the functional needs of the various packaging processes, and which can be verified simply and unambiguously. The main controlled dimensions are the height, horizontal body dimensions, verticality and, for cylindrical bottles, ovality; and most important of all, the bottle capacity.

HEIGHT

The main reason for good height control of bottles is so that they will make a good fit in their outer packaging, but undue variation may possibly affect some types of filling and capping operation. The overall height of a bottle is the vertical distance from the horizontal plane on which it is standing to the highest part of the finish. It is thus quite simple to make a fixed height gauge with cross-pieces corresponding to the maximum and minimum height permitted. An acceptable bottle will pass freely under the maximum gauge and fail to pass under the minimum one.

BODY DIMENSIONS

The main need for controlling horizontal body dimensions is to ensure that the bottles are not too large to pass smoothly and accurately through the guides, starwheels, etc., on the line. The ideal method of checking would be to construct a gauge in the same form as the most critical part of the filling line. For plain cylindrical bottles, a good approximation to this is obtained with a complete ring gauge made to correspond with the maximum acceptable bottle diameter. For cross-sections other than circular the gauges can be shaped accordingly, or else a fixed gap-gauge or dial calipers can be used across specified axes.

The minimum values of horizontal dimensions will rarely need to be checked by the packer. Overall bottle dimensions cannot fall significantly below the dimensions to which the mould is constructed; if a check is needed a ring gauge made to the minimum permitted dimensions would be sufficient. The only deviation which is likely to be measurable occurs when small local areas on a bottle surface are not fully blown up to the mould, or shrink away from it after blowing. There are some shapes of bottle, especially those with large flat panels, for which this warping cannot be completely prevented, and some bottles may be produced with dimensions generally within specification but with small patches below the minimum. It would be wrong to use a gauge which would throw out all such bottles on principle; this could easily happen if a gap gauge with very small gauge surfaces were used. The sensible thing to do is for the bottle maker and packer to decide together what size (and position) of sunken areas can be tolerated without harm, and to design any minimum gauge accordingly. It is possible to combine a ring gauge for the maximum diameter, with a suitably shaped gauge for the minimum, to make a single instrument. Fig. 6.12 is an illustration of a gauge of this type, taken from the British Standard BS 2812, for the finish dimensions of milk bottles.

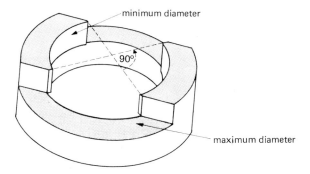

Figure 6.12.   Maximum and minimum gauge for milk

In special cases it may be that local bulges as well as sunken areas are objectionable for labelling reasons even though the dimensions are within specification throughout. If so, simple rocker gauges can be used to verify whether the amount of deviation across a labelling surface is acceptable or not.

OVALITY

Ovality was defined in Chapter 3 as the difference between maximum and minimum diameter in any horizontal cross-section of a cylindrical bottle.

It is not usually a very critical factor for packaging operations, unless the bottle is required to roll at some stage. It can only be verified with calipers by making two measurements at right angles and subtracting one from the other.

VERTICALITY

Verticality was defined as the horizontal distance by which the centre of the bottle finish deviates from its intended position in relation to the bottle base; this can be important for locating bottles accurately under filling or capping heads. Unfortunately there is no way of measuring verticality directly for all bottle shapes, but indirect measuring systems can be devised, working from reference points on the outer surface of the bottle, which when used carefully give a good practical estimate.

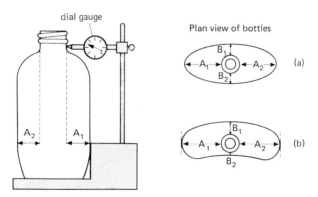

Figure 6.13. Verticality measurement

The principle of one method is illustrated in Fig. 6.13. The distances A_1, A_2, etc., shown in the plans are the *horizontal* distances between the side of the top of the bottle neck and the lowest vertical part of the body. (The latter point will usually be something like 25 mm above the base, avoiding the part where the body curves inwards towards the base.) For symmetrical bottles (Fig. 6.13a) the departure from verticality in the two directions is $\frac{1}{2}(A_1 - A_2)$ and $\frac{1}{2}(B_1 - B_2)$. For unsymmetrical bottles it is necessary to know or to measure the 'design values' of A_1, B_1, etc., and to calculate the deviation separately.

CAPACITY MEASUREMENT

Capacity is measured most simply and reliably by weighing the amount of water required to fill the internal space in the bottle, and we discussed in Chapter 3 how to fill the bottle accurately, either brimful or to the specified distance from the top. Bottle capacity is officially defined as its internal volume at the standard temperature of 20°C.

In the bad old pre-metric days the calculation needed was very simple because the fluid ounce was defined to be the volume occupied by one ounce of pure water, as weighed in air at 62°F (16·7°C). But the metric gram of water occupies slightly more than one millilitre, by an amount depending on its temperature. The following correction factors have been extracted from BS 1797 – *Tables for use in the calibration of volumetric glassware*.

Temperature of water and container (°C)	Volume occupied by 1 gram of water, when weighed in air (millilitres)
5	1·00149
10	1·00162
15	1·00208
20	1·00284
25	1·00387
30	1·00513

Thus the approximation that 1 ml of water weighs 1 g is not quite accurate enough, and a correction factor should be applied, usually amounting to between 0·2 and 0·4%. In practice the best and simplest method of measurement is first to weigh the bottle empty (and dry), then reweigh it after filling with clean water of known temperature.

Acceptance sampling schemes

Apart from the measurable characteristics of the packaging components, described statistically as 'variables', it is likely that there will be some defects present in the components, of different degrees of importance in relation to the components' performance. These are known as 'attributes', meaning that they are features which can only be recorded as being present or absent. The question arises as to whether the packer will find it useful to institute a formal system for inspecting random samples from every delivery, and deciding from the results whether to accept or reject the delivery. There seems to be no universal answer to this question as far as glass containers are concerned: a minority of packers do find it useful, but the majority do not. Quite a number have carried out trials, but have decided against continuation.

There is no shortage of advice about how to set up and conduct an acceptance sampling system, and the interested reader is referred to one or more of the following:

DEF–131A, *Sampling Procedures and Tables for Inspection by Attributes* (Ministry of Defence, 1965)

DG–7, *Guide to the Use of DEF–131A* (1966)

BS 6001, *Sampling Procedures and Tables for Inspection by Attributes* (British Standards Institution, 1972)

BS 6000, *Guide to the Use of BS 6001* (1972)

MIL–STD–105D, *Sampling Procedures and Tables for Inspection by Attributes* (US Department of Defense, 1963, amended 1964)

The steps in setting up a scheme are essentially the same in all cases:

1. The relevant defects should be defined exactly, if necessary by selecting and agreeing samples which show the degree of defectiveness.

2. If several defects are involved, they should be sorted into a small number of groups. Usually it is sufficient to have two main groups, one of Major defects which would be expected to affect the performance and one of Minor defects which are disliked for visual reasons. Critical defects, i.e. those which render the component unusable, are usually put in a special group of their own.

3. Decide on the percentage of defects in each group which it is thought would be just tolerable on average, known as the AQL (Acceptable Quality Level).

4. The tables provide for three types of sampling scheme: single, double and sequential; for glass containers single sampling is normally used. Choose from the relevant table a practicable sample size; the table will then indicate what the decision should be for any number of defects found.

It should be stressed that no scheme provides an infallible answer, especially for a single delivery of components. The aim is to provide a system for monitoring quality continuously and to remove some of the human bias from the making of decisions. In fact probably the most useful result of a scheme is that it can provide a continuous quantitative record which may show up any long-term trends in quality. And in some cases it may provide an increased chance of preventing trouble on the filling line.

On the other hand, it must be recognised that any scheme will cause a proportion of wrong decisions: of good deliveries being rejected and unacceptable ones passed; the smaller the sample size the greater the proportion of wrong decisions. The published tables show in graphical form exactly how great the risk is in each case. Further, the tables have been constructed on the assumption that the samples taken will be perfectly random, and anyone who has ever seen a lorry-load of bottles will know how impossible it is to achieve this requirement. Also in many plants there is just not time to take any samples at all before the load is used, and this is increasingly the case as packaging speeds increase.

Finally the cost of operating formal acceptance schemes is not insignificant, and those who have abandoned them have usually done so because the results obtained did not justify the cost. Also some have found that a false sense of security was being created, and the diligence of supervision on the filling line was suffering.

In fact the most reliable way of evaluating the quality of packaging components is to monitor their performance on the filling line. Even then, it is sometimes difficult separating the performance of the components from that of the packaging machinery itself, but it is always worth while making the effort. The best way if possible is to relate the performance records with the component manufacturer's own quality inspection records, when the genuine causes and effects can often be separated from the coincidences and unjustified assumptions.

7. Preparing bottles for filling

Whenever possible, bottles are transported direct from the bottle production line to the customer's premises, thus minimising handling and storage costs. Virtually all deliveries of bottles within Britain are now made by road transport rather than rail because of its lower cost, greater convenience and flexibility.

Direct delivery from the production line is possible only to customers who are able to cope with new bottles at about the same rate as they are being produced, i.e. between about 15,000 and 150,000 per day. In order to provide deliveries to other customers at the times and in the quantities required, bottle manufacturers have in the past maintained a network of warehouses at strategic points throughout the country, where stocks were held until required. However, because of the nation's improved motorway network, there is now a trend among the major glass manufacturers away from off-site warehousing in favour of holding stocks at or close to the point of production, and this is generally found to be more efficient and flexible than the previous methods.

The actual methods of packing bottles for transport have also changed considerably in recent years, with the virtual disappearance of returnable wooden cases. The main pack types now used are as follows.

1. RETURNABLE CORRUGATED FIBREBOARD CASES

Fibreboard cases strong enough to make several trips from the glassworks to the packing plant and back again are still used for handling bottles, but they are becoming less popular with the growth of alternative methods discussed below. The case dimensions and strength are chosen to suit the particular bottles, and usually the top and bottom flaps are sealed with gummed tape, thus providing a relatively dust-tight package. (For heavy bottles the bottom flaps may have to be stapled with metal stitches.)

The cases can be assembled on pallets and thus stacked to heights of 10 m or more in warehouses. Before returning for reuse, customers are asked to perform the simple operation of slitting the tape, collapsing the cases and making them into bundles.

In some situations the amount of reuse obtained does not justify the cost and effort incurred, and it may be found preferable to use a non-returnable case, probably made of lighter board.

2. CUSTOMERS' OWN FIBREBOARD CASES

In the past, the practice of using the same case or tray throughout the bottle's career – from glassworks to warehouse and thence to the filling plant, then on to distribution depot and retail store – was very widely followed, and is still fairly common.

For returnable bottles, it obviously simplified the packer's job to receive new bottles in the same pack as used for his returning empties. For non-returnable bottles, the packer was relieved of the need to co-ordinate deliveries of glass and fireboard, and he did not need to assemble the new cases, or return empty cases to the glass maker.

But disadvantages, too, have been found, particularly the loss of flexibility when a standard bottle needs to be filled with different products, or packed in cases marked to suit different outlets. And the spoilage or damage of some cases during the initial stages of storage and transport is not always insignificant. So times are changing, and more and more customers are finding it preferable to accept their glass in bulk-palletised loads, as described in the next section.

The design and construction of customer cases is discussed in Chapter 12.

3. PALLETS AND BULK PACKS

Wooden pallets are of course used extensively for storing and transporting packs of glass, as for most other mass-produced articles. With the arrival of metrication, UK glass manufacturers adopted a standard design of metric pallet, of dimensions and construction shown in Fig. 7.1. When pallet-loading cases or trays whose dimensions have been chosen to suit a particular bottle it is rare to find that they fit the pallet perfectly, and usually the only way to construct a stable load is to leave some spaces in the interior of the load pattern. This led to seeking ways of eliminating the wasteful gaps, and in the mid 1960s the first attempts were made to dispense entirely with the intermediate cases and pack the glass directly on to the pallet. Today the technique of bulk palletising has become established as the most important new method, offering significant benefits to both glass users and glass makers. Its success has depended to a

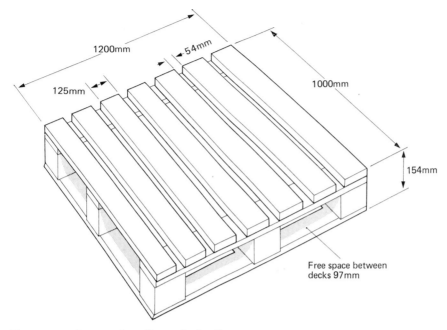

Figure 7.1.   Construction of a standard pallet

large extent on the development of shrink-wrapping techniques to pallet-size packs, as this system not only helps to obtain a secure load, but also provides good protection against dirt and the weather. It is also very easy for the packer to unwrap, either completely or partially as required. The film used is heavy-gauge polyethylene, with predetermined shrink characteristics.

A newer alternative to shrink-wrapping is that of stretch-wrapping, in which the pack is bound with a horizontally – or spirally – wound elastic 'bandage'. This has the potential advantage that the film tension is produced mechanically, so no shrink oven is needed. The transparent films used are not dissimilar to those available for domestic kitchen use.

In the simplest form of bulk pack, flat sheets of fibreboard the same size as the pallets are used to support each layer of bottles, hence called layer pads. The bottles are assembled closely together to cover as much of the pallet area as possible; sometimes with stable bottle designs it may even be possible to allow the outer rows to overhang slightly. Before shrink-wrapping, the pack is capped with an extra layer pad, or sometimes with an inverted fibreboard tray to give added stability.

If the bottle stability or the transport conditions leave something to be desired it may be advisable to stabilise the pack still more by using pallet-size trays for every layer instead of flat pads. The trays may be used upright, when a fairly shallow tray should be sufficient to steady the bases of the bottles standing in the tray. Alternatively the trays may be inverted over each layer of bottles, in which case the tray walls need to be deep enough to come down below the tapered neck of the bottles. (It is possible, but rarely necessary, to use 'belt and braces' by having an upright and an inverted tray for one or more layers.) A typical bulk pack with inverted trays is shown in Fig. 7.2.

Figure 7.2.   Bulk pack with inverted trays

If the layer pads or trays are made of solid chipboard, they can with advantage be returned and reused, but corrugated fibreboard can usually be used once only. Solid board trays can be made with slots and flaps for easy assembly and collapsing, without the need of gummed paper or other securing method.

There are two fundamental advantages of bulk pack. Firstly, it permits anything from 20 to 50% more bottles to be packed in a given volume, compared with the same bottle packed in conventional cases, and the savings in warehouse space and reduction of vehicle movements etc. can be very significant. As an example, a conventional cylindrical bottle of diameter 80·7 mm in customer's cases of a dozen packed 132 bottles to a pallet layer, compared with 203 bottles in a bulk pack layer, an improvement in area utilization of 54%. Secondly, the system lends itself to automatic load building in the glassworks, with a potential saving in labour, and similarly to automatic unloading at the filling plant, with similar savings and scope for very high-speed line feeding.

The system has the most obvious advantage to the packer who intends to pack his filled bottles into something other than the dual-purpose customer case, for example a shrink-wrapped tray, or one of the new types of 'wrap-around' cases. In many other situations, however, bulk packs may be found advantageous, and in a continuously developing situation packers would be well advised to investigate the system carefully with their glass suppliers. The costs of depalletising, either mechanically or manually, may well turn out to be more attractive than the costs of alternative systems.

As with all new developments, there are potential problems to overcome. One is the risk of damaging the surface of the bottles by abrasion in transit, but this can be minimised with good glass surface treatment. Another is the problem which can occasionally occur in damp climatic conditions of condensation taking place on the inside of the transparent film. There are several ways of coping with this if it is likely to be objectionable.

4. MODULAR TRAYS

For packers who are not yet ready to handle complete bulk packs, it is possible to divide each pallet layer into a number of sub-units which can be unloaded manually, using fibreboard trays (Fig. 7.3). If the tray dimensions are chosen so that an assembly of them will exactly fit the pallet, they are correctly described as 'modular trays'. Thus trays of external dimensions 600 × 400, 400 × 300, 400 × 200 mm, etc. will fill the 1200 × 1000 mm pallet exactly. Other dimensions can be found which are nearly as good, for example 500 × 333 and 333 × 250 mm will assemble into a solid layer of area of 1167 × 1000 mm. Such trays can be made either of corrugated board for single service, or solid board for reuse.

If the packer wishes to pack his filled bottles into shrink-wrapped shallow trays, there is the possibility of using the same tray both before and after filling. The trays could then be decorated or adapted in any way to suit the market requirements.

I

Figure 7.3.   Bulk pack with modular trays

5. OTHER METHODS OF DELIVERING EMPTY BOTTLES

Apart from the traditional shipwrecked mariner's method, many other systems have been and are being tried, and sometimes used commercially in specialised situations. We may refer briefly to the use of sealed paper sacks, wired or strapped bundles, and caged pallets in which the bottles are packed horizontally in a wire-mesh cage. It is also possible to shrink-wrap groups of bottles together without any other means of support. With bulk packs it is possible to build up a large closed fibreboard box round the pack, or alternatively to dispense with external protection altogether and to secure the pack with tensioned plastic strapping. But with the ever-increasing need to automate handling operations by means of commercially available machinery, such methods do not seem likely to become adopted widely.

Unpacking bottles from crates and fibreboard cases

Machinery is available for automatically dismantling stacks or pallet loads of bottle cases, but such equipment is usually built for one specific job (e.g. for unstacking milk bottle crates) and will not be discussed in detail. In this section we are concerned with methods for removing bottles from their crates or cases.

AUTOMATIC UNCASING

Some of the automatic machinery for taking bottles out of their cases is essentially similar to that used for repacking filled bottles. The latter operation is easier because repacking can usually be effected or assisted by gravity, whereas in unloading, each bottle is usually gripped individually and lifted upwards.

Automatic uncasing is mainly used on high-speed lines, particularly when large bottles are being handled. As a rough rule, automatic handling may be found economical when speeds exceed 100 bottles per minute. Much of the early de-crating equipment was designed primarily for the dairy industry, subsequently being adapted to suit beer and soft-drink bottles, and more recently, wine and spirit bottles.

As an example, we will consider the Vickers-Dawson de-crater, which in its 4-head version is capable of emptying 40 cases per minute. Fig. 7.4 shows a single-head de-crater unloading empty bottles and feeding them to a washing machine. The crates are automatically aligned under the head, which then descends pneumatically and grips the bottles by their necks; the bottles are lifted upward and forward, then deposited on the washer feed table. Changing to different sizes of bottle is a simple operation, because the heads slide off easily and the air connections have simple bayonet fittings. The gripping fingers are made of nylon to prevent damage to bottles or closures. When handling bottles in fibreboard cases, arrangements can be made to control the case flaps.

The Certus range of uncasers and repackers comes in a number of versions with 1 to 8 sets of gripper heads. The single-head units can handle up to 12 cases per minute, while the 8-head machines can deal with 63 cases per minute. In the single head unpacker, the whole case or crate is lifted up vertically to the gripper head; but machines with more than one head are equipped with a flight bar conveyor which spaces the cases correctly and positions them under the heads. The latter then descend and grip the bottles by their necks, using either inflatable rubber membranes which encircle the bottle necks, or mechanical grippers.

Figure 7.4. Vickers-Dawson single-head de-crater

A full list of uncasers and their manufacturers is given on pages 154–5. Many of them are similar in principle, varying mainly in the methods of gripping the bottles. Narrow-neck bottles are invariably gripped externally, but for wide-mouth jars various types of expanding components which fit inside individual jars can be used.

A different method of uncasing is employed in the New Way fibreboard case unloader (Fig. 7.5), supplied by Purdy. The bottom flaps are

Figure 7.5. New Way uncaser

first unsealed and the cases are then conveyed to the machine on a belt conveyor. The outer case flaps are automatically ploughed open and the weight of the contents opens the inner flaps. The containers are gradually lowered from the cases as they travel over a declining shoe. At the same time, two belts moving at an upward angle grip the sides of the cases and lift them away from the bottles, which remain on the moving conveyor for feeding to a unit which arranges them in single file. Output from this machine is up to 25 cases per minute.

SEMI-AUTOMATIC UNCASING

An alternative to fully automatic unpacking is to tip the bottles out of their cases by hand, and to use machinery for taking them away or marshalling them in single file. This method is used chiefly with fibreboard cases, and the filled cases should be sufficiently light and compact to be handled comfortably by the operators. The procedure is to open the case so that the bottoms are uppermost. (Returnable cases lend themselves well to this operation, as they can be opened at either end. For cases with stitched or glued bases, the bottles will have to be packed upside down by the bottle maker.) The open case is quickly inverted on to a table, preferably covered with a resilient material; then it is lifted off or used to guide the bottles forward on to a conveyor belt. The principle is thus similar to the operation performed automatically by the New Way unloader just described.

It can save effort and bottle abrasion if a tilting table is used for unloading; the open case is placed against the table when it is in a near vertical position, then the table is tilted into a horizontal position.

The Container-Feeder supplied by Pneumatic Scale is very useful for semi-automatic unpacking. With this machine the bottles are slid forward over tapered metal dividers which sort the bottles neatly into rows. The transporting belt moves forward intermittently, and with the aid of a stop-bar delivers orderly lines of bottles on to a single-line conveyor belt. The machine can deliver up to 30 lines of bottles, i.e. about 300 bottles, per minute, and as the main belt only moves intermittently there is very little jostling or abrasion of bottles.

Thomas Hill's Uniline single-file bottle feeder achieves the same result by a system of slat conveyors and transverse belts.

Under the heading of semi-automatic unpacking, mention should be made of the simple manually controlled de-craters supplied by Gimson and others, consisting of a suspended de-crating head which the operator guides into the mouth of the crate; compressed air does the rest.

MANUAL UNCASING

Manual uncasing includes all operations in which bottles are taken individually by hand out of their cases. The best way of doing this from a case without partitions is to place it at a convenient height, with its base at about 45° to the horizontal, and to take the bottles out starting at the top of the case. This is easier and safer than keeping the case horizontal, which may allow bottles to start falling about in the case after the first ones have been removed (an important point, often overlooked).

Whenever manual unpacking is being planned, it is wise to think particularly about the bottle necks and finishes. Careless or badly planned handling can cause damage here, which might not be detected until filling or capping.

Unpacking bottles from bulk packs

Not so long ago there were only two methods of transferring bottles from bulk packs to the filling line: slowly, by hand; or by means of high-speed automatic machinery operating at up to 900 containers a minute. Equipment for handling intermediate throughputs at realistic cost was completely lacking.

Today, however, the gap has been effectively filled, and several types of machinery are available to cater for different handling speeds.

In some filling plants, manual methods of depalletising may still be the best and cheapest, but as filling speeds rise above 120/minute, so mechanical systems increasingly gain the advantage.

AUTOMATIC DEPALLETISERS

There are two main types of automatic depalletiser: those with a 'sweep-off' action, and those with a 'lift-off' movement.

In the sweep-off machines, the pallet load is raised by means of a scissor-lift table or on cantilever forks until the top layer of ware is level with the filling line. A pneumatically, hydraulically or electrically operated sweep arm then pushes the complete layer of bottles on to a receiving area from where they are fed on to the filling line. Fig. 7.6 shows a Depallorator sweep-off depalletiser at Smedley–HP Foods, of Selby, Yorkshire.

The lift-off depalletisers are chiefly of value for handling tall or unstable bottles, which might tend to overbalance if sweep-type mechanisms were employed. They also enable upright pallet trays to be used more readily.

There are several methods for gripping and lifting the bottles from the pallets. One, used by Schaberger, employs a series of flexible air tubes which, when deflated, slip easily between the tapered necks of close-packed bottles. The tubes are then inflated by the machine's pneumatic system so as to bridge gaps of up to 65 mm between the necks of the bottles and exert sufficient pressure to enable the whole layer to be lifted gently, turned through a right angle, if required, and transferred on to an outfeed table for movement to the filling lines.

Another lifting method, used by Kettner, employs a platten with gripper rings which fit over the necks of the bottles. To cater for different

Figure 7.6. Depallorator sweep-off depalletiser

bottle sizes, the gripper head can be removed by disconnecting quick-release connectors and replaced by another gripper assembly.

Most of the gripping and lifting methods used in uncasers could also be adapted to depalletising. Some automatic depalletisers incorporate automatic equipment for the removal of the layer pads or pallet-size fibre-board trays. In others, this operation has to be done manually. The higher the line speed, the more profitable it becomes to employ automatic removal of layer pads or trays.

SEMI-AUTOMATIC DEPALLETISERS

Where only moderate filling speeds are employed, semi-automatic depalletisers can be used. Most of these machines simply raise the pallet load to the correct level for an operator to either remove the containers one at a time or transfer them on to a receiving area or unscrambling table by means of a simple sweeping action. Each layer of containers is raised and discharged in the same way until the pallet is empty. Through-puts of 100 to 150 per minute or more are obtainable with this method. The pallet lifting mechanism can consist of a hoist, a scissor-lift table or cantilever forks.

When planning this type of operation, it is useful to remember that it is not necessary to remove the whole of the shrink film at once. It is perfectly easy to cut it away partially with a knife, so as to free part of the load while holding the rest secure.

Methods of cleaning used bottles

In considering methods of cleaning bottles, we must distinguish between used returnable bottles, which will certainly need a thorough washing, and new bottles, which sometimes need little or no treatment at all. We will consider first the conventional washing process, which is intended primarily for cleaning returned multi-trip bottles.

WASHING MACHINERY

There is a surprisingly long list of washing machines (see end of chapter), and it would take a complete book to describe all their features of design and operation. Here we can only discuss the general principles.

Modern washers are divided into two main groups: the straight 'hydro' types, in which all the washing is done by internal and external liquid jets, and the 'soaker-hydro' types, in which part of the washing is done by total immersion. In a few of the latter, mechanical brushes are used instead of some of the jets. There is scope for endless discussion about which group is the better, but undoubtedly both are capable of excellent results, and the fact that most manufacturers make both speaks for itself. There are some situations where one or the other system is indicated; for example, bottles with hard sugary deposits will benefit from a soak, while bottles which are nearly clean need only a brief rinse with jets. Most returnable bottles – milks, beers and carbonated minerals – can be washed either way.

The throughputs of both types of washer have increased tremendously over the past decade, to match – or indeed surpass – the increase in bottle filling speeds. Some soaker-hydro machines are now capable of washing up to 2000 bottles per minute compared with a maximum of about 600 ten years ago. Similarly, the speed of the largest modern hydro-jetting machines has doubled during the same period from about 400 to 800 per minute or more.

METHODS OF BOTTLE MOVEMENT

There are many designs of bottle holder. Hydro-washers require only a short funnel-shaped holder, to carry the inverted bottles over the jets, while holders for soaker washers need to be larger, to hold the bottles firmly at all angles required. It is good practice for the parts which contact the bottles to be made of rubber, nylon or polypropylene, but care is needed in the choice of material, because some plastics do not stand up well to the prolonged action of hot alkaline detergents. Usually the holders are arranged in rows, which are linked together to form a wide endless belt which travels round inside the washer. In some designs of holder, the bottle necks protrude slightly through the bottom. The depth of protrusion depends on the shape of the bottle shoulder, and some packers have been surprised, when trying to wash a slim bottle in a washer designed for a broad one, to find that the bottles emerge without their necks.

Until fairly recently most bottle washers employed the principle of intermittent motion (i.e. the rows of bottles moved through the machine intermittently at the rate of perhaps 6 or 10 steps forward per minute). But some of the latest machines – especially those designed for high through-puts – employ continuously moving conveyors which carry the bottles non-stop through the sequence of operations. Besides making for more gentle handling of the bottles, these provide savings in power consumption and cut down on wear and tear of the moving parts.

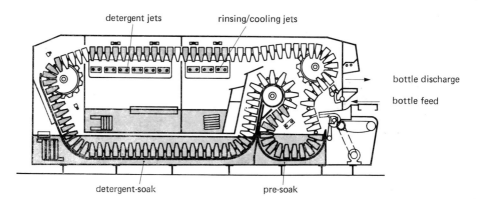

Figure 7.7.   Cross-section through Udec L15 soaker-hydro washer

Most large washers work on the 'straight-through' principle, with bottles entering at one end and leaving at the other. Small and medium ones are often of the 'come-back' type, which feed and discharge at the same end; this arrangement may help to save space. Come-back washers have to be of the soaker-hydro type, because the bottles will spend part of the time inverted in the top half of the washer, and the rest of the time upright in the lower half. Fig. 7.7 shows a vertical section through a Udec washer of the come-back type. A general view of the same machine

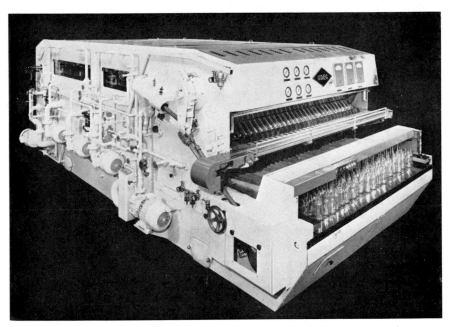

Figure 7.8. General view of Udec L15 soaker-hydro washer (U.D. Engineering)

is seen in Fig. 7.8. In this particular machine the belt moves forward 15 times a minute, propelled hydraulically, so in the version which has a belt 28 bottles wide, the output can reach 420 bottles a minute. In machines of this general type higher outputs can be obtained by increasing the width of the belt to take more bottles, or by increasing the length and adding extra soaking and jetting stages.

A different system of movement is used in the cup chain hydro machines made by British Miller Hydro and others. In these the bottles travel inverted in single file through the machine, and the carrying chain winds round a number of horizontal sprockets, each of which contain moving jets which synchronise with the moving bottles. The arrangement is very versatile, and can be tailored to suit special requirements; for example, with stages of steam sterilisation or hot air drying. Automatic bottle loading and unloading can be arranged with this type of washer, but they are also used very widely with hand loading. Nowadays these machines are employed mainly by spirits bottlers for general-purpose lines handling a wide range of bottle shapes and sizes.

Gimson washers have an unusual method of operation which enables good use to be made of limited space. They consist of a row of coaxial drums, and the bottles make a complete revolution in each drum in turn, soaking the lower half and being treated with jets in the other half.

A new Vickers-Dawson continuous-motion soaker-hydro, described as a single-ended machine, in fact offers an original approach to washer design in that both 'ends' are located in the middle of the machine, facing each other at the same level. The advantage of having feed and discharge at the same point, as in a conventional 'come-back' machine, is thereby retained, and both functions can be supervised by the same operator.

In ordinary single-ended washers, the incoming bottles descend downwards into the soaker tank, but in this machine they first travel upwards to pre-rinse jets. Then they move down into the soaking tank, which extends the full length of the machine, running under a walkway used by the operator at the feed and discharge point. What in an ordinary machine would be floor space is thus used as bottle-cleaning space. Detergent jetting and rinsing with fresh water are done by banks of jets which move in synchronisation with the continuous motion of the bottle carriers, and then return. The machine is suitable for beer, soft drink and milk bottles, and is available in a range of sizes to handle outputs of up to 1400 per minute or more.

CHOICE OF DETERGENT

Plain water is not very good for bottle washing, and even if the washing is only intended to remove loose dust or dirt, a mild detergent or at least a wetting agent will improve the efficiency and reliability of the process. The traditional bottle detergent is caustic soda, used in concentrations from about 0·5 to 3%, and many packers believe that this is still the best. Many others prefer the 'balanced' detergents, which are mixtures of caustic soda with other detergent chemicals, e.g. sodium silicates, carbonates and phosphates, designed to wash as efficiently as possible at concentrations of about 0·5 to 1%.

There is, of course, no simple answer to what is the best type of detergent, but washing-machine manufacturers will be able to advise what has been found most suitable for their particular machines and the various types of bottle. The main jobs the detergent may have to perform are:

Wetting. The detergent must thoroughly wet the glass surface and also any solid matter which can be dissolved. Nearly all detergent chemicals achieve this easily; in fact there is a danger of overdoing the wetting action, causing excessive foam in the washer, particularly in the jetting sections.

Emulsification and suspension. Greasy deposits must be broken up into small globules which can be detached from the glass (emulsification), and other solid matter must be broken into small particles which will go into and stay in the liquid (suspension). Caustic soda does this quite well, but some of the silicate/phosphate combinations are better.

Chemical destruction. Fats and greases can be converted by caustic soda into water-soluble soaps (hydrolysis). Other chemicals can only do this in proportion to their caustic alkalinity. Occasionally these chemical effects can be overdone, for example, if paper labels are reduced to a clogging pulp.

*Sterilisation.** The germ-killing power of detergent chemicals usually depends on their caustic alkalinity, so caustic soda is better than most other chemicals. A notable exception to this rule is the range of quaternary ammonium compounds, which have outstanding bactericidal power.

Rinsing. Once the bottle is clean, it must be rinsed free of detergent without recontamination by dirt. This means that the detergent must maintain its emulsifying and suspending powers even while it is being greatly diluted by the rinse water. Caustic soda is moderately good, but some silicates and phosphates are better. The outside of the bottle, as well as the inside, must be free of detergent, otherwise difficulties could occur in labelling.

There are three other points, not directly concerned with the cleanliness of the bottle:

Washer lubrication. It is quite an engineering problem to haul perhaps several tons of bottles and carriers through hot detergent, without excessive friction or mechanical wear. Caustic soda is a useful lubricant in this respect, and some machines cannot work without it. Silicates and phos-

---

* In the context of this book the word sterilisation is used in a commercial rather than an absolute sense, to describe any process designed to destroy micro-organisms. It should not be assumed, therefore, that a 'sterilised' container or material necessarily reaches the standard of sterility needed in, say, a surgical operation.

phates have the opposite effect, and if used indiscriminately can even make some machines seize up completely.

Scale prevention.   Hard water is liable to cause scale on metal parts. Caustic soda does not prevent this, but several other chemicals can control it in various ways. Prevention, however, is better than cure, and the use of soft water is to be recommended.

Effect on glass surfaces.   The effect of detergents on glass surfaces is not significant unless the surface has first been damaged by prolonged abrasion; then visible marks may develop ('scuffing') which have sometimes been troublesome with long-lived bottles. The author has found no difference in the effect of various possible detergents used on milk bottles; if scuffing starts to occur, the remedy is to control the physical damage.*

CONTROL OF WASHING TEMPERATURES

Heat is an essential adjunct to detergents in bottle washing, but heat alone is no substitute for detergents. Temperatures in commercial washers rarely need to exceed 80°C, and usually a maximum of about 63°C is found to be adequate.

The maximum sudden temperature changes which various types of bottle can withstand have been discussed in Chapter 4: in practice washers are usually adjusted so that the temperature changes are about half the bottle's limiting values. The recommended temperatures for a typical washer when washing milk bottles could be as follows:

	°C	Time of contract
Pre-soak	35	1 min
Soak	63	3 min
Detergent jets	63	10 sec
Hot rinse	43	20 sec
Warm rinse	35	20–30 sec
Cold rinse	18	10 sec

Time is important as well as temperature. If, for example, the hot-rinse stage given above were reduced to five seconds, the effective bottle temperature would still be somewhere near 60°C when the bottles passed to

* See 'The Scuffing of Milk Bottles', by B. E. Moody. *Journal of Society of Dairy Technology*, **12**, 1959, p. 15.

the warm rinse at 35°C; this might be near the borderline of safety for bottles which were worn or slightly damaged after long service.

A little care may be needed in extremely cold weather; if the initial bottle temperatures are below freezing point, it might be desirable to reduce slightly the temperature of the first warm soak or spray.

CONDITION OF BOTTLES AFTER WASHING

Usually bottles emerge from the washer cold and wet after their last rinse, and this is the most convenient condition if they are to be filled immediately, with milk, carbonated beverages or most other cold aqueous liquids. If the product is to be filled very hot, for example jam, the final rinse should be made with hot water, so as to lessen any thermal shocks.

Dry products, of course, need to be filled into dry bottles, and there are many other pharmaceutical, chemical and food products for which a wet bottle would be unsuitable. Also if some time elapses between washing and filling, it may be necessary to dry the bottles. This is done by incorporating a hot-air drying stage in the washer. When bottles are dried in this way, a little extra care may be needed in the subsequent handling, because bottles are more sensitive to surface damage when they are hot, clean and dry, than at any other time.

Treatment of new bottles before filling

Many products can be filled into new bottles without any prior bottle treatment, provided the bottles have been handled with reasonable care, and kept in closed cases or bulk packs until the time of use. For most foods, drinks and pharmaceutical preparations, however, some type of pre-treatment is desirable, if only as a safety precaution.

At least four different ways of cleaning new bottles are possible:

1. They can be washed in exactly the same way as for used bottles. This is obviously the simplest method for new *returnable* bottles, which are being fed into a filling line designed primarily for used bottles.

2. They can be washed briefly with detergent, or rinsed with water, in a hydro-washer. Only a short contact time will be needed, and a simple machine, through which bottles pass continuously in single file, will be satisfactory in many instances.

For example, British Miller Hydro's Mains Water Rinser has been designed specifically for cleaning non-returnable bottles. Polypropylene pockets convey the bottles in a single line and invert them around a multi-spoke wheel, during which time they are jetted with mains pressure cold water. The bottles continue in an inverted position through a long drain-ing section before being re-inverted around a wheel at the other end of the machine. Throughputs of up to 500 bottles per minute are possible.

Other machines designed for cleaning one-trip bottles resemble the conventional multi-lane hydro bottle washers, but are shorter in length and incorporate fewer jetting stations. The 'Superpower R' range, made by Vickers-Dawson, for example, can be built in a number of versions varying from a plain three-stage high-pressure cold-water rinser to equipment incorporating detergent washing, pre-rinsing and final rinsing. Outputs range up to 1000 bottles per minute.

3. If an absolutely sterile or aseptic container is needed, as, for example, when packing certain antibiotics, a severe treatment will be needed, involving the use of dry heat or high-pressure steam. An alternative possibility is the use of ethylene oxide. This gas is extremely lethal, even in low concentration, to all organisms, and all that is required is a method for ensuring that the gas penetrates right inside the container. The lethal nature of ethylene oxide applies also to human beings, and very strict precautions are needed when handling it.

4. The air cleaning of new bottles is simple, cheap and can be very effective if carried out properly. As this subject is becoming of ever increasing importance, it will be discussed in some detail.

Air cleaning

Air cleaning is not just another method of cleaning dirty bottles, to be considered as an alternative to washing with detergents; it is the final stage in a process designed to preserve the initial cleanliness and sterility of the new bottle right up to the time of filling. Most of the other stages have to be carried out during and immediately after the manufacture of the bottles, so a decision to use air cleaning must be communicated to the bottle maker and all other persons involved, before the bottles are made, and possibly even before bottle design begins.

When the newly made bottle leaves the mould in which it was blown, its interior is at a temperature of several hundred degrees, so it is com-pletely sterile. While it cools in the annealing lehr a little hot air enters the bottle, and experience has shown that this air is remarkably free of bacteria or similar contamination. So if the bottle could be filled imme-diately after annealing, no cleaning whatsoever would be necessary.

Bottles being made for air cleaning are either bulk palletised and shrink-wrapped or packed immediately into fibreboard cases, which are then sealed with gummed tape, or securely closed by some other method.

A simple method of sealing cases is to fold the flaps down, and to stack them upside-down on pallets, except for the bottom layer of cases which is placed the right way up. This method is very effective provided the cases remain on the same pallet throughout their storage and transport. The bottles should be delivered and used as soon as possible, but while awaiting use they should be stored under clean conditions and not exposed to excessive variations of atmospheric temperature and humidity. (If the period of storage runs into several months, air cleaning may not be wholly effective, unless the atmospheric conditions have been unusually stable. In this case a conventional liquid wash would be needed.)

If the storage conditions are such that there is a risk of condensation on the glass, this can largely be circumvented by bringing the bottles out of the storage space into the air cleaning area 24 hours before use.

When unpacked for filling, the bottles should be as clean as when they were made, so much so that several packers have found experimentally that a liquid wash at this stage is likely to decrease the sterility rather than to improve it. The essential point is that whereas a bottle may be micro-biologically clean immediately after washing, it can only remain as clean as the atmosphere in which it cools. Consequently, the conditions of bottles which have been packed at the lehr end, and properly stored and transported, is better than can be achieved with bottles washed and dried in a packaging plant.

The bottles may, however, contain a little loose dust or fibre, acquired during transport. The sole purpose of air cleaning is to remove this loose dust and any other foreign bodies. It is done by blowing a jet of *dry filtered* compressed air into the bottles, to sweep the inside surface clear of all small particles of dust and fibre. (The air must be effectively filtered of moisture, oil and solid matter.)

AIR CLEANING MACHINERY

In all commercial air cleaners, the basic principle of blowing dry filtered air into the bottle is used. For narrow-neck bottles it is essential that the blowing nozzle extends well inside the bottle, and this is usually desirable also for wide-mouth jars. In some machines the movement of air inside the bottle is assisted by applying suction at the bottle mouth at the same time as blowing inside.

It may be mentioned here that some bottle designs are not suitable for air cleaning; for example, those with very angular shoulders, very narrow necks or with complex body shapes.

Most machines perform the operation with the bottle inverted, for

K

obvious reasons, but there are some suitable for treating shallow jars in an upright position. The methods of inverting the bottles range from the simple manual process of placing a bottle upside-down over a jet, to various ways of swinging the bottle round in a circle. In one type the line of bottles is led into a drum which rotates through 180° about its horizontal axis, presenting a line of inverted bottles to air nozzles waiting to rise into the bottle necks. In other types the bottles travel continuously in holders which form the spokes of a large vertical wheel. If the wheel is placed above the main conveyor belt, the bottles travel with their necks towards the centre of the wheel, and if it is placed below, the necks point outwards.

Most drum and vertical wheel machines can provide outputs of 120 bottles per minute or more, depending on the number of blowing heads. For example, the Thomas Hill vertically rotating Airline machine Mk II can be supplied with 9, 12, 18, 30 or 40 blowing spindles for outputs of up to 600 bottles per minute.

At the other end of the scale, there are simple manually operated air-cleaning machines capable of 20 to 50 bottles per minute. These do a good job if properly used, but it must be pointed out that a machine in which

Figure 7.9. Kemwall semi-automatic air cleaner

the blowing time is left to the operator's discretion will produce more variable results than one in which the time is fixed by the action of the machine.

A full list of air-cleaning equipment is given at the end of the chapter, but a few examples are discussed more fully below to illustrate the general range.

Fig. 7.9 shows a simple semi-automatic machine made by the Kemwall Engineering Co. There are 16 blowing spindles on the rotating table, and the blowing section extends for 60° of the circle. Loading and unloading is by hand, but the blowing time is determined by the machine speed. Ejected particles fall by gravity into a removable bag.

In the Banks Sanitair machine, made by Morgan Fairest, the bottles are carried in a horizontal circular path. Each carrier holds two bottles in an opposed position (i.e. base-to-base) which halves the amount of space that would otherwise be needed. On entering the machine (seen in Fig. 7.10) the bottle is inverted, allowing large heavy particles to fall out. An air tube is then raised into the top of the inverted container and a blast of compressed air is released. Vacuum is then applied, drawing any foreign matter into a removable bag. Each container is in the machine for $1\frac{1}{2}$ revolutions. Outputs of 500 bottles per minute can be obtained.

Figure 7.10 Banks Sanitair air cleaner

The Gilowy multi-lane air cleaners are different from most other automatic air cleaners in that they do not require in-line feeding. Bottles can be emptied in bulk from fibreboard cases or automatically depalletised on to a broad conveyor which channels them into multi-lane bottle inverting and blowing stations. Various models are available for handling different types and sizes of bottles and jars at up to 300 per minute.

Figure 7.11.   Pneumaclean air cleaner

Finally, Fig. 7.11 shows the high-speed Pneumaclean machine made by Pneumatic Scale, which is also of interest for its unusual method of conveying the bottles. The conveyor belt leading into the machine takes the bottles between two endless inflated tyres, made of Neoprene, which converge to grip the bottles and carry them through the rest of the machine smoothly and noiselessly. The cleaning action is achieved partly by centrifugal force as the bottles go round the driving wheel, and partly by ten stationary jets along the bottom of the machine. This flexible method of conveyance makes possible quite complex and very fast manoeuvres in a small space – the whole action of the Pneumaclean is contained between floor level and the standard conveyor height of one metre.

Inspection of bottles before filling

It is often useful to provide a station just before the filler where the bottles can be inspected visually. This provides a final check that there are no foreign bodies in any bottles, and that none are chipped or damaged when they reach the filler. With returnable bottles it is necessary to check also that all the returned bottles are of the correct design, and this can be done most conveniently after washing, when the bottles are clean and lined up accessibly. Good eyesight and an illuminated translucent screen behind the conveyor line are needed for this operation.

Reasonably good inspection can be achieved by one person at speeds of about 60 bottles per minute, but the operators will need relieving at fairly frequent intervals if efficiency is to be maintained. At very high speeds, visual inspection can be carried out by splitting the lines of bottles into two or three slower lines, and recombining them after inspection. Some fast-filling machines are specially designed to allow this recombination.

BOTTLE SORTING EQUIPMENT

Some fully automatic equipment is available to assist the sorting of returned empty bottles, and more is under development. Barry-Weh-miller, for example, supply equipment which can be used to pick out any bottles of incorrect height and diameter. It will also segregate coloured from white flint bottles and reject any which bear a plastic or metal closure.

Incoming bottles are fed to a starwheel at up to 600 bottles per minute and then passed between the front and rear lobes of an inspection head. Here, each bottle in turn is inspected by four optical sensing systems. The first senses the position of the bottle and activates the electronic circuits of the other three sensors. The second is a colour sensor, the third a dimension sensor, and the fourth detects the presence of closures.

The mode of operation of each sensor (e.g. over-height bottles passed, coloured bottles rejected, etc.) can be selected by means of a micro-switch. Insertion of a programme card enables different combinations of these functions to be chosen.

Rejected bottles are taken out of the main bottling line by a starwheel fitted with rubber sucker cups at each bottle position. When the inspection head detects an unwanted bottle, a vacuum is applied to the sucker cup, and the bottle is carried round by the starwheel on to a reject conveyor or accumulating table. Accepted bottles pass on to the main bottling line.

BOTTLE INSPECTION EQUIPMENT

Automatic inspection of bottles for contamination or foreign bodies is fairly simple if the bodies are large and opaque, but some difficulty may be encountered if the sensitivity has to be increased too far.

Several manufacturers supply equipment which operates by passing a beam of light upwards through the base of the bottle and into an inspection head containing an optical scanning system and an arrangement of photocells. Usually, the bottles are conveyed into the inspection position by means of a rotary turret.

Most of the machines are similar in principle, but their optical and electronic systems vary. Normally in order to detect the shadow caused by a small object one has to move it rapidly past the detector or scan it with a moving light beam in order to get a fluctuating signal. With these machines, however, the same effect is produced very neatly with effectively a static light beam and a static bottle by spinning the optical system above the bottle, thereby producing a rapidly moving image of the bottle base (including any contaminating particles inside it).

In the Barry-Wehmiller Opti-Scan machine, for example, the light rays from the base of the container are first made parallel by a lens, then passed through a rotating prism. On emerging, the rays are brought into focus by another lens so as to produce a rotating image of the bottle base on a bank of photocells. Anything inside the base of the container will cast a shadow as part of the moving image. The consequent change in cell illumination produces an electrical impulse which is amplified and used to operate a pneumatic reject system similar to that described earlier for rotary bottle sorters. There are four photocells in all, of which three are sensitive only to the rapid variations in light caused by small foreign objects. The fourth photocell is sensitive to longer-term changes in the light level, due, say, to a coloured bottle or one which is grossly contaminated. This photocell signals the reject system if the mean light intensity falls below a predetermined limit. The machine is capable of dealing with up to 800 bottles per minute.

The Fords (Finsbury) bottle inspector (Fig. 7.12) works on a rather different principle as the electronic inspection head has no moving parts. There is no rotary turret, but bottles travel into the inspection position on a diffuser disc which is free to rotate. A small 'head' of bottles on the infeed conveyor is sufficient to establish the flow. At the inspection position light is directed from the base of the bottle on to a solid-state device, which monitors the intensity of light received by 208 diodes, each scanning a separate area. Small objects are detected by comparing the outputs for each area with the mean light intensity as measured by another diode. Any reduction in the level of light received by one or more of the 208 scanning diodes automatically triggers the reject mechanism. The mean-level diode output is, in turn, compared with a reference time unit to detect a completely obscured bottle base or mouth. Inspection is complete after the bottle has moved only 3 mm. Rejected bottles are

Figure 7.12. Fords automatic bottle inspector

pushed from the inspection position into a collection tray by a pneumatic ram, while accepted bottles continue along the conveyor.

In all base inspection equipment a compromise has to be reached between acceptance and rejection. If the sensitivity of the equipment is set too high, some bottles may be rejected which could have been filled. For example, bottles from different sources might have different types and degrees of embossing on the base, and this could interfere with the scanning system. Interference should not be a problem if the embossed markings are well rounded and correctly positioned, but in any case the bottle bases can be left completely clear if this is specified in advance.

In Japan, a machine which combines base inspection with sidewall inspection has been developed by Mitsubishi. This can not only detect the presence of foreign bodies, but also detects broken bottles and chipped sealing surfaces. Sidewall inspection is by means of a probe carrying a complex optical system which is lowered into the bottle while the container is revolving. The probe inspects the inside and outside of the bottle and can detect objects of about 5 × 5 mm in size. Chipped or cracked sealing surfaces are detected by applying compressed air. An unsound sealing surface will cause a pressure drop, and this is detected by a pressure switch which operates the rejection system.

Another bottle inspector, which combines base inspection with mouth inspection, has been developed by Datz of West Germany, and is available in the UK through Crown Cork. This is a rotary machine which detects cracks and other flaws in the mouth of the bottle by directing light rays on the mouth from several different directions. The plates on which the bottles are carried in the rotary turret are fitted with photocells which detect any fault in the base of the bottle, or contamination inside it. The machine can be fitted with 8 to 16 heads.

List of machinery for preparing bottles for filling

In this and later machinery lists the name of the British supplier is given; if the machine is not normally built in the UK the name of the manufacturer is added in brackets. Machine speeds depend on many factors, such as size and shape of bottle or case and type of product. The maximum speeds quoted here have been taken directly from the suppliers' literature, and often they are not for exactly comparable conditions. They should, therefore, be regarded only as a rough guide to possible performance.

Bottle and jar depalletisers

Supplier	Machine name	Principle of operation*	Max. speed per min.
Bosch (Strunck)	Various	Bottles lifted as needed S/A and A. Also powered sweep	2 to 3 layers
		Manual or powered sweep	4 layers
Conpack Engineering (Leifeld & Lemke)	Fleximat Brüser	S/A three types Powered sweep	— 1200 bottles
Emhart (Standard-Knapp)	—	Manual or powered sweep	1400 bottles
Erben (Kettner)	—	Bottles lifted by grippers	600 bottles
Glenhove (NIKO)	—	Powered sweep on to elevated conveyor forming part of M/C	600 bottles
Goddard (H. Schaefer KG)	Schaefer	Manual or powered sweep	1500 bottles
Holland	—	Manual or powered sweep	3 layers
Mather & Platt (Standard-Knapp)	—	Manual or powered sweep	1400 bottles
MetaMatic (Metal Box)	—	Powered sweep S/A and A	1000 bottles
Pantin	—	Manual or powered sweep. S/A and A	1000 bottles
Penrose	Pallet-sweep	Manual or powered sweep. S/A and A	900 bottles
Platt	Reynolds-Platt	Manual or powered sweep. S/A and A	—
Power Lifts	—	Manual or powered sweep	2 to 3 layers

* Abbreviations used for principle of operation: S/A Semi-automatic; A Automatic.

Bottle and jar depalletisers (continued)

Supplier	Machine name	Principle of operation	Max. speed per min.
Rockwell Packaging Machines (Schaberger)	—	Bottles lifted by inflatable tubes S/A and A. Also powered sweep	—
Stork-Amsterdam	Stork	Manual or powered sweep	1500 bottles
Van den Bergh Walkinshaw (OCME and Depallorator)	—	Manual or powered sweep	1500 bottles

De-craters and unpackers

Supplier	No. of heads	Max. no. of cases per min.	Notes
Andrew & Suter (ABC)	—	25	Lifts fibreboard cases off bottles
Bosch (Strunck)	1 to 3	36	Heads descend to grip bottles or hold bottles and lower fibreboard tray
Bramigk (Simonazzi)	1 to 8	93	Heads descend to grip bottles
Conpack Engineering (Leifeld & Lemke)	1 to 6	7 to 50	Heads descend to grip bottles
	Double unit	14 to 80	Heads descend to grip bottles
Crown Cork (Baele)	2 to 12	100	Heads descend to grip bottles
Erben	1 to 8	63	Heads descend to grip bottles
Fairest Trading (Certus)	1 to 8	75	Heads descend to grip bottles
Gimson	1	—	Manually guided and operated

Glenhove (Datz)	1 to 4	41	Heads descend to grip bottles
Holstein & Kappert	Various	85	Heads descend to grip bottles
Meadowcroft	1 to 4	58	Heads descend to grip bottles
Meyer	3 to 10	50	
Purdy (New Way)	—	25	Lifts fibreboard cases off bottles
Remy	3 to 10	75	Heads move round carousel, rising and falling to pick up and set down bottles
Rockwell Pneumatic Scale (Pneumatic Scale)	—	300	Semi-automatic belt-type feeder
Stork-Amsterdam	1 to 4	60	Heads descend to grip bottles
U.D. Engineering	1 and 2	17 and 30	Case lifted to decrater head
Van den Bergh Walkinshaw	Various	50	Heads descend to grip bottles (and unscrew if require
Vickers-Dawson	1 to 8	72	Heads descend to grip bottles

Washers

Supplier	Names	Principle*	Max. no. of bottles per min.	Notes
Adelphi	—	Hy, S/A	—	Single head, for medicals
Anglo Continental (Prot)	Various	So/Hy and Hy	660	For vials
(Neri & Bonotto)	New Zenith and Super Zenith	—	100	For vials and small bottles

* Abbreviations used for principle of operation: So Soaker; Hy Hydro; So/Hy Soaker/Hydro; S/A Semi-automatic (speeds are not quoted for these).

Washers (continued)

Supplier	Names	Principle	Max. no. of bottles per min.	Notes
Anglo Continental (CIONI)	C10 and C12	—	300	For vials and ampoules
Autopack	Ampak	Hy	300	Ampoules
Barry-Wehmiller	Hydro-Jet	So/Hy	1500	
Bosch (Strunck)	Various	Hy and ultrasonic	600	For vials, ampoules and bottles
Bramigk (Simonazzi)	Elba	So/Hy	333	—
Cockx	—	Hy, S/A	—	Twin head, with brushes
Crown Cork	Crown-Ladewig	Hy and So/Hy	1000	Double-ended and single-ended
	Baele	So/Hy	1430	Double-ended
	Baele	So/Hy	900	Single-ended
Emhart (Standard-Knapp)	Aqua-Air	Hy/Air	600	Twist-type, single file
	367 Rinser	Hy	1200	Single file
Engelmann & Buckham	'Stone'	Hy and So/Hy	—	
Erben	Various	So/Hy	930	Double-ended and single-ended
Gimson	Junior, Mechanical	So/Hy	120 & 200	Rotating drum principle
	Hydraulic Mk. VI	So/Hy	300	
Glenhove (NIKO)	—	Hy	250	Single file
	—	Hy	333	Double track
Goddard (Kronseder)	Contistella	So/Hy	—	Single-ended
Hart (Cozzoli)	Various	Hy, So/Hy	300	Ampoules and small bottles
Hill	Speke Mark III	Hy	180	Inverted vial rinsing

Hill (continued)	Grangemouth	Hy, S/A	50	
	Cleanline	Hy	300	Twist-type single file for vials, jars and bottles
Holstein & Kappert	Omega-Conti	So/Hy	1670	Single-ended
Manning	Various	Hy	100	Ampoules and small vials
Mather & Platt (Standard-Knapp)	Aqua-Air	Hy	—	Twist-type single file
	367 Rinser	Hy	1200	Single file
Meadowcroft	Meadowcroft	So/Hy	300	Single-ended
MetaMatic (Metal Box)	SNR–1	Hy	600	Air/water rinse followed by drain-off
Meyer	Dyna-guard	So/Hy	1000	
Miller	Junior	Hy	60	
	C, BPH, Best-ever and Pint-type	Hy	48, 60, 70 & 400	Single file
	Straight-through	Hy	600	
	50 and 400	So/Hy	600	
Pacunion (Bausch & Ströbel)	Various	Hy/Air	116	Small bottles and ampoules
		So/Hy	200	Ditto, with ultrasonic washing
		Hy, S/A	—	—
Stork-Amsterdam	MF/MFL MFLS/MR	So/Hy	900	—
U.D. Engineering	S5	So/Hy	50–100	Single-ended
	S6	So/Hy	80–200	Single-ended
	S7	So/Hy	80–150	Single-ended
	L15	So/Hy	150–450	Single-ended
	Straight-Through Type 5	Hy	200–800	—
Van den Bergh Walkinshaw	B & M	So/Hy	60–600	
Vickers-Dawson	Little Gem	Hy	35	Rotary

Washers (continued)

Supplier	Names	Principle	Max. no. of bottles per min.	Notes
Vickers-Dawson (continued)	HM	Hy	120	Single-ended
	Vertex, Apex	Hy	12, 38	Rotary
	Cyclops	Hy	240	Double-ended
	New Junior	Hy	120	Double-ended
	Superpower H	Hy	600	Double-ended
	Superpower R	Hy	1000	Double-ended
	Dominant	Hy	1000	Double-ended
	Defiant	Hy	200	Double-ended
	Dominus	So/Hy	300	Single-ended
	Durus	So/Hy	310	Single-ended
	New single-ended	So/Hy	1400	Single-ended
	Superpower S	So/Hy	2000	Double-ended

Air cleaners

Supplier	Names	Principle	Max. no. of bottles per min.	Notes*
Anglo Continental (Zanasi Nigris)	SU65	Single-file	180	Small bottles
Bosch (Strunck)	RTS A 04 RTL A 01	S/A and A single file	300	
Conpack Engineering (Leifeld & Lemke)	Brüser system	Continuous single file, A	240	Bottles and jars
Erben (Gilowy)	—	Multi-lane	300	
Flexile	Farmomac F36	Single file, continuous	120	Bottles
	Coster PR 611	Single file, continuous	300	Bottles and aerosols
Hill	Duet	2-head, S/A	—	
	Rotary Air	48 rotating heads, S/A	—	Bottles only

* Bottles here means narrow-neck containers, as opposed to wide-mouth jars.

Hill (continued)	Watford	Continuous, single-file	150	Jars in upright position
	Airline	9 to 40 heads, vertical wheel	600	
	Cleanline	Continuous, single file	300	Jars
	Miniclean	20 and 30 rotating heads	300	Vials and small bottles
	Miraclean	Continuous, single file	600	Uses air or water
Kemwall	—	2-head, S/A	—	
	—	16 rotating heads, S/A	—	
King	RB3	12 rotating heads	90 and 150	
	BBS	Static in-line 1 to 6 heads	120	
Morgan Fairest	Sanitair	Duplex rotating heads	500	
Raglen Packaging	Newmapak	Single file	250	
Rockwell	—	2-head, S/A	—	
Pneumatic Scale (Pneumatic Scale)	—	Automatic inverted drum type	120	
	Rotacleaner	12-head, vertical wheel	120	Bottles only
	Ultracleaner	18-head, horizontal wheel	300	
	Pneumaclean	Rubber tyre conveyor	500	
Van den Bergh Walkinshaw	B & M	Single file	300	Rinsing possible

Bottle sorters

Supplier	Machine name	Principle	Max. input of bottles per min.	Notes
Barry-Wehmiller	—	Electronic	600	Single reject channel
Emhart	Autosort	—	—	—
Flexile	Farmomac F31	Rotary	300	—
	Coster CV 000	Rotary	300	—
Glenhove (Datz)	—	Electronic	666	Single reject channel
Goddard (Kronseder)	Krones Inspektor	Electronic	600	Single reject channel

Bottle inspectors

Supplier	Machine name	Principle of operation	Max. input of bottles per min.
Barry-Wehmiller	Opti-scan	Electronic	800
Crown Cork (Datz)	—	Electronic	600
	—	Electronic	1000
Engelmann & Buckham	Scop-E	Electronic	—
Fords (Finsbury)	—	Electronic	400
Glenhove (Datz)	—	Electronic	666
Goddard (Kronseder)	Krones Inspektor	Electronic	750
Meyer	Sentinel	Electronic	600

8. Filling

Liquid and semi-liquid filling

Filling bottles with liquid by means of a pipe from a supply tank sounds a simple operation, and indeed it is so; the only real problem is deciding how to operate the tap. Many of the large filling machines in use today, however, appear to be very complicated, and it is quite impossible just by looking at them to see how they work. The published information is also often rather confusing – perhaps some of it is written to impress rather than to inform the reader – and some of the technical terms are not clearly explained, or are used differently by different people.

Part of the reason for the confusion may be that some writers try to explain simultaneously the basic principles of a machine and the engineering details of its construction. In this chapter an attempt will be made to avoid this pitfall, by discussing filling in the following order:

1. The various basic principles of operation.
2. The practical requirements of commercial fillers.
3. Examples of machines of each type available in the UK, and a discussion of their particular purposes and abilities.

Basic principles of fillers

Liquid will flow through a pipe system only if there is a higher pressure at one end than the other. Theoretically the rate of flow, through a cylindrical pipe, is given by Poiseuille's equation:*

$$V = \frac{\pi p r^4}{8\eta l}$$

where

V = volume flowing per second
p = pressure difference causing the flow†
r = internal radius of pipe
η = viscosity of the liquid
l = length of pipe

* This holds accurately only for conditions of streamline flow, and low rates of flow. More complicated empirical equations have to be used for fast or turbulent flow, some of which contain $\sqrt{p}$ in place of p.

† In this chapter, pressure differences are given either in terms of the head of water which produces them, or else in the number of bars above or below atmospheric pressure.

L

Most fillers make use of gravity to cause the flow, and so are called *gravity fillers*. Fig. 8.1 shows four elementary gravity fillers, each consisting of an open tank with a pipe coming out of the bottom; the relative rates of flow calculated from Poiseuille's equation are indicated in the diagram.

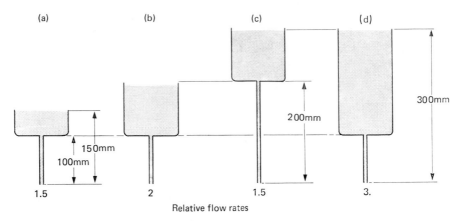

Figure 8.1.   Elementary gravity fillers

It would probably surprise most plumbers to realise that (c) gives only the same flow rate as (a), even though the gravity head is twice as high: the effect of the greater pressure is cancelled out because the length of pipe in (c) is twice that in (a). The practical point is that flow rates are governed by the narrowest section of the pipe system, and so larger gravity heads do not produce faster flow if they are obtained by lengthening a section of narrow pipe.

In Fig. 8.1 atmospheric pressure has no effect on the flow, because the air pressure is effectively the same at both ends of the system. If, however, the top of the tank is closed, and the air pressure inside the tank raised, the flow rate will be increased, because the total driving pressure is now the sum of the extra air pressure in the tank, plus the gravity head as before. A machine working on this principle is strictly a *pressure-gravity filler*. In practice, if the air pressure is much greater than the gravity head, the effect of the latter can be ignored, and the arrangement called simply a *pressure filler*.

If, on the other hand, the lower end of the pipe is connected to a vessel in which the pressure is less than atmospheric, the pressure difference in the filler is again increased; such a filler is called a *vacuum filler* (or strictly a *gravity-vacuum filler*).

In practice we must distinguish between six types of gravity filler, two types of pressure filler and two of vacuum filler. In addition there are three miscellaneous classes:

Gravity	*Pressure*
Siphon	Simple pressure
Simple gravity	'Snifted pressure'
'Vacuum-assisted' gravity	
Direct counter-pressure	*Vacuum*
Differential counter-pressure (one stage)	Low vacuum (or fixed vacuum)
Differential counter-pressure (two stage)	Adjustable vacuum

Miscellaneous

Combinations of gravity, pressure and vacuum
Predetermined fill
Automatically controlled fill

The principles on which each of these systems work will now be explained and illustrated. In every system there is a means for maintaining a constant height of liquid in the supply tank, usually a ball valve or something similar; for simplicity this has been omitted from all the diagrams. Shut-off valves are shown diagrammatically:

OPEN

CLOSED

In most diagrams a small graph is included to show how the effective pressure-head (i.e. the amount of pressure which makes the liquid flow) changes during the time of filling of the bottle. Most systems are illustrated before filling, during filling and after filling is complete; the three stages are indicated in the diagrams by the letters A, B and C.

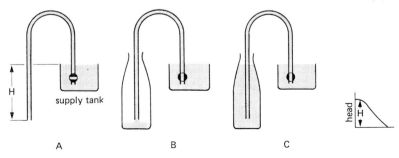

Figure 8.2. Siphon filler

SIPHON FILLERS

The principle of the siphon is well known. Liquid will start to flow as soon as the valve is opened, even if no bottle is there, and will continue until the level in the bottle is the same as that in the tank. The available pressure head is initially H, and remains constant until the end of the filling tube is submerged, then decreases steadily to zero as the fill continues. (Figure 8.2.).

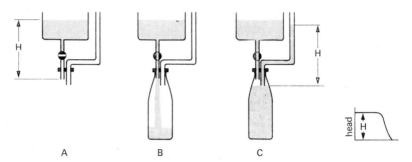

A   B   C

Figure 8.3.   Simple gravity filler

SIMPLE GRAVITY FILLERS

The bottle is placed in position and the valve opened. Liquid will flow whether or not the bottle makes an airtight seal with the filling head; if the seal is correct the flow will cease almost as soon as the bottom of the air tube is submerged. If the valve is left open, equilibrium will be reached when the pressure in the headspace is H above atmospheric. With the simple system illustrated, there would be a practical difficulty, because liquid would run back from the air tube when the bottle was removed.

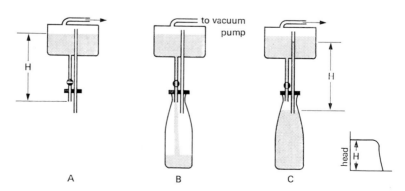

A   B   C

Figure 8.4.   'Vacuum-assisted' gravity filler

This could be cured by leading this tube out sideways to some form of overflow vessel, but in this case flow would continue until the filling valve was closed.

'VACUUM-ASSISTED' GRAVITY FILLERS

This method overcomes the problem of run-back from the outlet tube. The supply tank is closed and the air space in it maintained at a slightly reduced pressure, usually about 0·05 bars below atmospheric, equivalent to a negative pressure-head of 0·5 m of water. When the bottle is in position, making an airtight seal against the filler, the pressure in the bottle rapidly falls to the same as that in the supply tank. When the valve opens, liquid flows *by gravity* into the bottle, under a gravity-head H. Filling ceases very soon after the outlet tube is submerged, when the liquid in this tube has risen to a height H above the level in the bottle. At this stage the pressure in the bottle headspace will be slightly higher (by an amount H) than the air pressure in the supply tank.

In practice the liquid near the top of the bottle may be bubbly or frothy, and therefore less dense than the rest of the liquid. In this case the return flow up the outlet tube will not stop at the level shown in the diagram, and the froth can be pulled right up into the supply tank. When the filled bottle is removed, all the liquid remaining in the outlet tube can be pulled up into the supply tank. It is not customary to have a valve on the outlet pipe, so air is continually sucked up through this pipe when no bottle is in position.

It should be noted that in this system the use of vacuum does not affect the flow rate, compared with that in a simple gravity filler. The main purpose of using vacuum is to avoid filling bottles with damaged sealing surfaces, and to overcome the problem of dripping when the bottle is removed.

DIRECT COUNTER-PRESSURE FILLERS

Carbonated liquids cannot be handled in vacuum systems, nor usually at atmospheric pressure, as there would be excessive foaming and loss of carbon dioxide. The 'direct counter-pressure' filler is a gravity filler which works on exactly the same principle as the 'vacuum-assisted' one, except the the air space in the supply tank is maintained at a high pressure, in order to keep the dissolved CO_2 in solution throughout the process. Also a pressure valve is added to the outlet pipe to prevent loss of pressure when no bottle is in position. Sometimes the filling tube is extended down to near the bottom of the bottle, in order to promote quiet filling. This does not affect the principle, except that it slightly alters the gravity-head available for filling, as shown in Fig. 8.5.

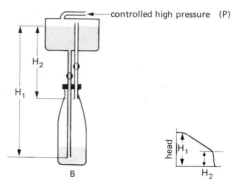

Figure 8.5.   Direct counter-pressure filler

When the bottle is placed in position, the pressure valve is opened first, and when the full pressure has been established in the bottle the liquid valve is opened. When filling stops, the pressure in the headspace is $P + H_2$, and it is usually necessary to provide a 'snift valve' in the filling head (not shown), to release this pressure to the atmosphere before the bottle is removed. If, however, the volume of the long filling tube is such that it takes up all the space required for the vacuity in the capped bottle, the bottle will be brimful at the end of the filling cycle and there will be no gas to release through a snift valve.

In some direct counter-pressure fillers the gravity head, H, may be as much as 0·7 m, and in these cases the overall pressure needs to be fairly high, of the order of 3–4 bars,* to prevent excessive evolution of CO_2. Such fillers are therefore sometimes called *high-pressure* fillers. Other fillers working on exactly the same principle have gravity heads as low as 0·15 m, and the overall pressure then can be reduced to around 2 bars; sometimes these are described as *low-pressure* fillers. However, there is no clear dividing line between high and low pressure, and both terms are better avoided when describing counter-pressure fillers.

DIFFERENTIAL COUNTER-PRESSURE FILLERS: ONE-STAGE

In this system the effective gravity head is reduced by making the pressure in the bottle initially somewhat greater than that in the supply tank. By

---

* These pressures, of course, depend on the degree of carbonation of the liquid being handled.

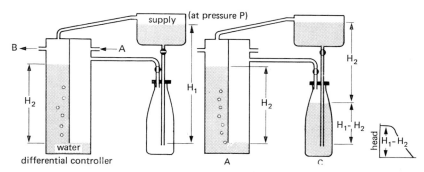

Figure 8.6. One-stage differential counter-pressure filler

this means the carbonated liquid will flow more quietly into the bottle, and consequently the whole system may be operated at a lower pressure than in a direct counter-pressure system. Hence the names *low-pressure filler* or *balanced* pressure filler are sometimes used.

The pressure controller consists of a vessel with a dividing wall as shown. The controller is partly filled with water, and compressed air slowly admitted at A, forcing the water in the right-hand compartment down to the bottom of the dividing wall, until air starts to bubble under the wall to the left-hand side. At B there is a regulating valve which maintains the pressure in the compartment and in the supply tank at the required value, which we will call P. This means that so long as air is flowing through the controller, the pressure in the right-hand side will be $P + H_2$. (The pressure of the compressed air supply must of course be slightly greater than $P + H_2$.)

When the bottle is placed in position, a pressure of $P + H_2$ is established inside it. When the liquid valve opens, there is available a pressure-head of $P + H_1$, so the net filling pressure at the start is $(P + H_1) - (P + H_2) = H_1 - H_2$. This pressure decreases as the liquid level in the bottle rises, and flow stops when the liquid in the bottle has risen to within H_2 of the level in the tank; the pressure in the headspace is still $P + H_2$, and this may be released through a snift valve. During filling, air in the bottle passes out through the controller to the regulating valve.

In practice, H_1 may be of the order of 0·7 m, and H_2 0·5 m, making the initial head available for filling 0·2 m. Obviously the filler will not work at all unless H_1 is greater than H_2, so the filling tube must extend below the required filling level in the bottle.

The volume of the differential controller needs to be much larger than that of the bottle, in order to avoid fluctuations in pressures during filling.

DIFFERENTIAL COUNTER-PRESSURE FILLERS: TWO-STAGE

The two-stage system is exactly similar to the single-stage one in principle, but the effective filling pressure is increased to a higher value after the end of the filling tube is submerged, when there is less danger of frothing. In the first stage of filling, the bottle is connected to the right-hand compartment of the controller, by means of valve C, and so is at a pressure of $P + H_2$. Filling starts under a net head of $H_1 - H_2$, and would stop when H_0 decreased to become equal to H_2. Just before this happens, valve C is closed and valve D opened; for the rest of the filling cycle the pressure in the bottle is maintained at $P + H_3$, and so the last stage of filling continues at a faster rate, until H_0 decreases to become equal to H_3. In practice, $H_1 - H_2$ may be of the order of 0·1 m, and the initial head in the second stage about 0·15 m.

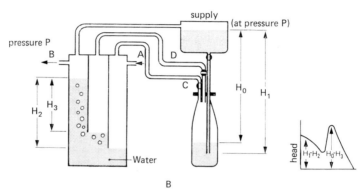

Figure 8.7.   Two-stage differential counter-pressure filler

As with the other types of counter-pressure filler, the final pressure in the headspace ($P + H_3$) is usually released through a snift valve before the bottle is removed.

SIMPLE PRESSURE FILLERS

The principle of a simple pressure filler is exactly the same as that illustrated in Fig. 8.3 for a simple gravity filler, except that the top of the supply tank is closed, and maintained at the required high pressure, P, by a compressed air supply.

When the bottle is full, the pressure would be capable of forcing the excess liquid to a height $P + H$ above the level in the bottle, if the bottle were sealed to the filling head. In practice an additional valve would be put in the outlet pipe and closed to prevent overflow, alternatively the pipe would be led to some form of overflow vessel.

'SNIFTED PRESSURE' FILLERS

This is a particular form of pressure filler for carbonated liquids in which the valve in the outlet pipe, the snift valve, is opened intermittently during the filling cycle. The principle is straightforward. The bottle is placed in position with the snift valve closed, and the liquid valve is opened. Liquid is forced in by the high pressure, until the pressure in the bottle equals that in the supply tank. The initial pressure in the bottle is 1 bar, so if the tank pressure is 4 bars the air or gas in the bottle will be compressed to roughly one-quarter of its original volume when the flow stops; thus the bottle will be about three-quarters full of liquid. At this stage the snift valve is opened gently and briefly, to release some or all of the excess pressure; filling can then continue. This sequence may be repeated several times, until the required volume of liquid is obtained in the bottle. The snift valve is then opened once more before the bottle is removed.

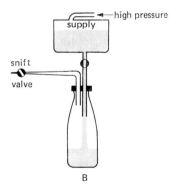

Figure 8.8. Snifted pressure filler

The expression 'open snift filler' is occasionally used to describe a multi-head filler in which each head has an individually adjusted snift valve. The more usual and better arrangement is the 'controlled snift' filler, in which all the snift valves are led to a central manifold, and controlled collectively.

LOW-VACUUM FILLERS

Low-vacuum fillers (sometimes called fixed-vacuum fillers) are much used for still liquids, i.e. any liquids which are not liable to bubble or froth excessively when the pressure above them is reduced.

The filler is connected to a vacuum pump or exhauster, capable of maintaining a constant pressure slightly below atmospheric, so that the

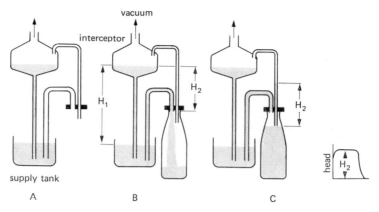

Figure 8.9   Low-vacuum filler

liquid in the central column will be maintained at a height H_1 (of the order of 0·7 m) above the constant level in the supply tank. Thus if the atmospheric pressure is P, the pressure in the interceptor vessel is $P - H_1$. When the bottle is sealed in position against the filler, the pressure in the bottle also quickly becomes $P - H_1$ as air is sucked out to the interceptor, and filling commences with a pressure head which quickly increases to a value H_2. Filling stops when the liquid in the interceptor tube has risen to a height H_2 above that in the bottle. (It is a common misconception of low-vacuum fillers that when the bottle is full the liquid continues to circulate up through the tube to the interceptor and down through the central column. This is clearly not possible, as it would be 'perpetual motion'. It is, however, possible to lift froth from the surface of the bottle into the interceptor.)

At the end of filling, the pressure in the bottle is $P - H_1 + H_2$, i.e. quite close to atmospheric pressure.

It will be noted that the system is capable of operation without any valves. There should nevertheless be no dripping when the bottle is removed, because the vacuum can then pull the liquid in the interceptor tube up into the interceptor, and the liquid in the filling tube returns to the position shown in Fig. 8.9.(A).

ADJUSTABLE-VACUUM FILLERS

In adjustable-vacuum fillers the degree of vacuum used can be altered to suit the conditions required, without affecting the principle of operation.

This system differs from the low-vacuum one in that flow does continue into the interceptor when the bottle is full. Usually the flow is stopped by opening another valve in the filling head (not shown) which

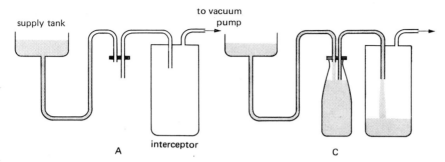

Figure 8.10   Adjustable-vacuum filler

allows the bottle headspace to return to atmospheric pressure. Devices are included to return the liquid which accumulates in the interceptor to the main supply tank.

COMBINATIONS OF PRESSURE, GRAVITY AND VACUUM

Fillers can be constructed using any two, or all three, of the possible driving forces. Machines also exist in which the method of filling can be altered as required, for example, from vacuum filling to pressure filling.

PREDETERMINED FILL

In all the methods discussed so far, the bottle plays an important part in determining how much liquid goes into it. Another possibility is for the filling machine itself to do the whole job of measuring out the required quantity.

Usually this is done by volume; liquid is first measured into a cup or cylinder of known volume, and then all of it is discharged into the bottle. Movement of the liquid may be achieved by air pressure, gravity or vacuum, and extra pressure may be provided by the movement of a piston.

Another method for the pressure filling of viscous liquids is simply to pump liquid into the bottle at a constant rate for a constant time, but this is not strictly a 'predetermined' fill. Finally, direct weight control is possible, but for liquids it would have no advantage over the methods already mentioned.

AUTOMATICALLY CONTROLLED FILL

There are several other methods of stopping the liquid flow as soon as the bottle is sufficiently full. One machine makes use of a fine air jet to detect when the filling height has reached the required position; the filling tube is then automatically withdrawn. Electrical contact switches can be used in the same way, and beams of light or X-rays can be set to make an interruptor system.*

Practical considerations in liquid filler design

Any automatic bottle filler has to meet three basic requirements:

1. It must operate at the required speed.
2. It must be able to place the bottles correctly in the filling position, and remove them after filling.
3. It must provide the means for starting and stopping the flow of liquid as needed.

FILLING SPEEDS

Of the three requirements listed, the first is usually the most demanding, and has the greatest effect on the machine design and cost, so it should be considered first. We have already seen that the rate of outflow from a filling tube depends on the diameter of the tube, the pressure-head and the viscosity. The machine designer cannot do much to alter any of these: the tube must be small enough to go smoothly into the bottle neck, leaving room for one or possibly two outlet tubes; the properties of the liquid probably fix a limit to what pressure-head can be used; and the viscosity is fixed (unless it can be reduced by raising the filling temperature). Quite a lot can be done, however, to improve smoothness and speed of flow, by good design of the various pipes, nozzles and valves, by avoiding constrictions and sudden corners or bends, and especially by ensuring that all internal surfaces are as smooth as possible.

When all these points have been taken care of, the situation may be that a certain bottle can be filled and levelled off in, say, six seconds, but not less. A single filling head could thus fill perhaps eight bottles per minute, allowing some non-filling time for placing and removing the bottles. If a filling speed of, say, 320 bottles/min is required, the filling machine will need to have 320/8 = 40 heads. (In practice the calculation is not quite so simple as this, because the proportion of non-filling time to filling time varies with the number of heads, the size of bottle and other factors.) High-speed machines are usually of the rotary type, so that they can run continuously, and as many as 164 heads have been provided, producing filling speeds up to 1670 bottles per minute. Straight-line fillers have a

*See also the discussion of fill-level detectors in Chapter 10.

row of stationary filling heads positioned over a single-file bottle conveyor; higher speeds can be obtained with a double row of heads.

Greater filling speeds are possible when a machine is designed to fill only one size of bottle than when a general-purpose machine is demanded capable of filling a range of different sizes. For example, there is a wide choice of general-purpose machines for filling beer and minerals, and their maximum speeds for small bottles ($\frac{1}{4}$ to $\frac{1}{3}$ litre) lie in the range 3 to 5 bottles per filling head per minute. If, however, a machine is designed for these small bottles alone, speeds of up to about 10 bottles per head per minute have been obtained.

BOTTLE HANDLING ON FILLERS

Placing bottles in position for filling is relatively easy. Most automatic fillers place the bottle on a pedestal which is lifted up to the filling head; the valve assembly usually carries guides or centring cones which ensure that the bottle neck reaches the right position. The pedestals are variously moved by cams, springs or air cylinders; whichever method is used, it is important that the up-and-down movement should be well controlled and not jerky. Air pressure can provide very smooth control, but if leaks occur in the system, the smoothness is soon lost. A cam action is very simple, but the bearing surfaces must be hard-wearing and well lubricated, and the thrust should not put an excessive strain on the main machine bearings. Perhaps the best methods are those which combine some form of pneumatic and mechanical action.

DESIGN OF FILLING VALVES

The heart of the filling machine is the filling valve. Many air and liquid valves are operated externally by cams or levers on the body of the machine. This makes the 'plumbing' simple, but it is necessary to have additional devices to ensure that the filling head does not operate if a bottle is missing or unsuitable for filling. Alternatively the valve can be opened by the movement of the bottle itself; this is done with most milk fillers, but it is still necessary to have a method of preventing defective bottles being filled. Another method of valve operation, much used in counter-pressure fillers, is to open the air valve mechanically, and to operate the liquid valve by the change of pressure in the bottle. Examples will be shown in the following sections.

The length of the filling tube and the position of the valve also vary. Some machines have a long filling tube descending almost to the bottom of the bottle, in which case the valve may be placed right at the bottom of the tube, in the form of an inverted cone or 'pear'. The tube is closed by lifting the pear with a rod running down the centre of the tube. This is excellent for preventing drips, but even a fine wire along the axis of a tube reduces the maximum flow rate. Short filling tubes usually have their valves outside the bottle neck.

As mentioned earlier, the low-vacuum machine needs no valves at all, and this has several advantages in construction and operation.

CONTROL OF FILLING HEIGHT

It is a common but incorrect idea that all fillers (except the volumetric ones) fill to the exact height of the bottom of the outlet tube. It is true that in most of the systems described so far the filling height is determined by the position of this tube, but in all cases the final liquid level will be slightly higher than the bottom of the tube. This is partly because of the momentum of the liquid in the flow pipe, but also because the flow does not stop until the air trapped in the headspace has been compressed to a pressure sufficient to balance the head causing flow. For example, in Fig. 8.5 the gas in the headspace is theoretically at a pressure of P at the moment when the bottom of the outer tube is sealed off by liquid, but flow does not stop until this gas has been compressed to a pressure of $P + H_2$. In practice there are several reasons why some variation in filling height occurs, for example, a little liquid may sometimes drip from the liquid tube while the bottle is being withdrawn.

Some fillers provide an extra stage of fine adjustment to improve filling height control. In one method the bottle is first intentionally overfilled, then the excess is sucked off. To do this the bottle is slightly withdrawn from the filling head, keeping the end of the outlet tube under the liquid surface; at the same time the filling valve is closed. The excess liquid can then be sucked off to just below the level of the end of the outlet tube.

If the final adjustment is made with the bottle out of contact with the filling head, as shown in Fig. 8.11, it will be seen that the height of fill obtained, H, is equal to the distance between the surface of the bottle support and the bottom of the outlet tube.

In most cases the bottle obviously plays a major part in determining the contents volume. In fact there are several different ways in which this may take place:

1. In some of the simple systems described earlier, the bottle is filled to a 'fixed' distance from the top, determined by the distance that the outlet tube projects into the bottle neck, less the small amount of over-

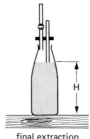

filling completed　　　　　final extraction

Figure 8.11.　Adjustment of filling height from bottom of bottle

shoot or other error. (Here the length of the *filling* tube is quite irrelevant!) In other fillers the fill height is determined indirectly, by means of the various pressure heads in the system. So in these cases the bottle content is largely determined by its capacity at a height from the top corresponding to the intended filling point.

2. With the extra stage of 'suck-off' just described, the bottle ends up filled to a more or less fixed distance measured from the bottom of the bottle.

3. When the filling tube projects below the liquid level in the bottle, as happens with some counter-pressure fillers, the level will obviously fall when the tube is withdrawn. The characteristics of the bottle which affects the final content is its capacity at the fill height *before* the tube is withdrawn. In the extreme case this may be the brimful capacity.

4. With snifted pressure fillers the brimful capacity is always the controlling characteristic.

Thus we can see the connection with the EEC Measuring Container Directive (discussed in Chapter 5), which specify two types of measuring container related to two types of filling. The Measuring Container with its capacity controlled at a fixed height from its top is supposed to correspond to cases 1 and 2 above, and the one with controlled brimful capacity to cases 3 and 4. The theory is that with Measuring Containers specified and controlled in this way, the packer and the Government Inspector can virtually check the bottle contents by simply watching the filling machine. We can see that with some types of filling operation this is a rather major over-simplification; apart from the complications already discussed, no account is taken of the fact that many products are filled hot or very cold, so the packer has to set his initial fill height high or low in order to get the 'correct' one when the product temperature falls or rises to 20°C. A theoretically more satisfactory system would have been to specify the Measuring Container's capacity at the height which the packer has decided he has to aim for in the bottle at the end of the filling cycle, just before the filling head is removed from the bottle. In reality, of course, this would be very complicated, and the over-simplified system should work quite well. Most measuring containers will probably have their capacities specified at a height from the top, and anyone will be able to check filled containers at any time after filling with relative ease. The only fallacy is to suggest that this is a fundamental change, from control at 'point of sale' to control at 'point of manufacture'.

Practical applications and examples

The practical applications of each type of filler will now be considered, and examples of commercial machines described in detail. The relative costs of machines will not be discussed, but they are obviously important when deciding between a simple and elaborate machine. A comprehensive list of machines and suppliers is given at the end of the chapter, together with the method of operation, number of heads and filling speeds.

SIPHON FILLERS

Siphon fillers were the only multi-head machines available until the mid-nineteenth century,* and were widely used for still liquids and for beer. As far as the author is aware, very few if any survive in the UK, but the principle is still employed in rotary and straight-line fillers used in other parts of Europe.

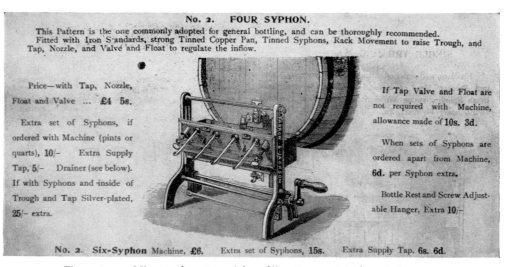

No. 2.   FOUR SYPHON.

This Pattern is the one commonly adopted for general bottling, and can be thoroughly recommended. Fitted with Iron Standards, strong Tinned Copper Pan, Tinned Syphons, Rack Movement to raise Trough, and Tap, Nozzle, and Valve and Float to regulate the inflow.

Price—with Tap, Nozzle, Float and Valve ... £4 5s.

Extra set of Syphons, if ordered with Machine (pints or quarts), 10/-   Extra Supply Tap, 5/-   Drainer (see below). If with Syphons and inside of Trough and Tap Silver-plated, 25/- extra.

If Tap Valve and Float are not required with Machine, allowance made of 10s. 3d.

When sets of Syphons are ordered apart from Machine, 6d. per Syphon extra.

Bottle Rest and Screw Adjustable Hanger, Extra 10/-

No. 2. Six-Syphon Machine, **£6.**   Extra set of Syphons, **15s.**   Extra Supply Tap, **6s. 6d.**

Figure 8.12.   Nineteenth century siphon filler, Farrow & Jackson Ltd.

* The earliest record of a siphon filler is a patent granted to Thomas Masterman in 1825.

As this type is thus largely a matter of history, the illustration chosen, Fig. 8.12, is of a 'general purpose' filler in use from perhaps the middle of the nineteenth century. It is taken from an old catalogue of Farrow & Jackson Ltd (established 1798).

The siphon tubes were pivoted at the top, with counterweights to raise the open ends when not in action. A valve at the submerged end opened automatically when the open end was pulled down. The filler was primed by sucking.

SIMPLE GRAVITY FILLERS

These are generally very similar in appearance and use to the 'vacuum-assisted' gravity fillers, so they do not need a separate illustration.

'VACUUM-ASSISTED' GRAVITY FILLERS

This is one of the methods particularly suitable for milk, as it combines the possibility of high-speed filling with the simplicity of design essential for really thorough machine cleaning. Modern machines can fill up to 10 pint milk bottles per filling head per minute.

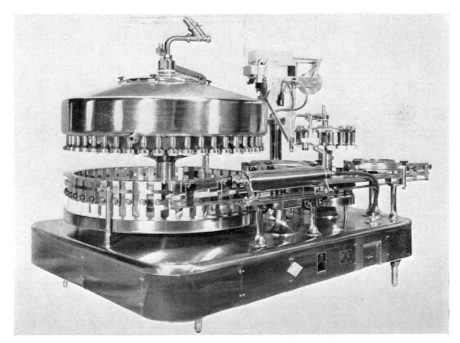

Figure 8.13.   A 48-head Udec vacuum-assisted gravity filler (U.D. Engineering)

M

In the Udec machine, shown in Fig. 8.13, the bottles are delivered by screw feed and starwheel on to the bottle lifts. The lifts are held by springs in the raised position, and are pulled down for loading and unloading by a roller running under a cam track. As the bottle is raised to the filling position its neck opens a rubber valve. The machine has an integral capping unit as is normal with milk fillers.

When filling milk under reduced pressure, some air becomes entrapped in the stream of liquid, especially if it does not flow smoothly down the sides of the bottle. Time must therefore be allowed for the air bubbles to rise to the surface and be drawn off before the bottle is removed from the filler. In order to obtain good control of filling levels, it is usual to withdraw the bottle slightly from the valve (or vice versa) when it is full, so that the milk supply stops and the final levelling-off can be done accurately.

DIRECT COUNTER-PRESSURE FILLERS

Direct counter-pressure fillers are widely used for high-speed filling of beer, cider and carbonated minerals. The example shown in Fig. 8.14 is the Super Cemco 60-head beer filler, supplied in the UK by Crown Cork.

The bottles are led on to bottle lifts, which are raised by compressed air and lowered by cam action. A feature of high-speed machines of this type is a method for combining bottles from two separate conveyors into a single screw feed, running at 500 or more bottles per minute.

The Cemco filling valve has two air connections to the supply tank and one liquid connection, and is controlled by an external lever which can be turned to three positions. The sequence of operations is as follows:

a. The bottle neck is lifted into the centring bell and sealed against the valve. The lever is turned to position 1, which opens both air connections, admitting counter pressure to the bottle. The beer valve remains closed.

b. The lever is moved by a cam to position 2, closing one of the two air connections, and opening the beer valve. Beer flows into the bottle until it is filled to the top of the neck; the time required for this is between one-half and three-quarters of the machine revolution.

c. The lever is moved to position 3, closing all the valves.

d. The bottle is withdrawn from the filling head.

In the filling position, the filling tube reaches nearly to the bottom of the bottle, and the withdrawal of the tube after filling provides the necessary vacuity in the bottle.

As with nearly all fillers for carbonated liquids, an integral capper is provided to apply the crowns; the Cemco 60 has 15 capping heads.

Figure 8.14.   Super Cemco-60 direct counter-pressure filler

DIFFERENTIAL COUNTER-PRESSURE FILLERS

Although there are many examples of one-stage and two-stage differential counter-pressure fillers still in operation in this country, machines of this type are no longer manufactured commercially, chiefly because they cannot equal the higher speeds afforded by direct counter-pressure fillers. They do, however, give a quiet, efficient fill, and two examples typical of the machines which may still be seen in operation are selected for inclusion here.

Fig. 8.15.a shows a close-up view of bottles entering and leaving a Worssam Super-speed 60-head beer filler. This is a one-stage differential

Figure 8.15a. Worssam Super-speed Sixty one-stage differential filler

counter-pressure filler and the method of operation can best be followed by comparing Fig. 8.15.b with the schematic diagram in Fig. 8.6. The supply tank is placed on top of the differential controller, and the dividing wall between the two compartments of the controller is a cylinder, inside the outer wall of the controller. The filling head is automatically operated by compressed air, and the only action which has to be performed mechanically is the opening and closing of the air supply valve at the beginning and end of the filling cycle. The filling tube extends to the centre of the bottle, and is controlled by a pear valve at the lower end. Adjustment of the filling level can be carried out very simply by altering the amount of water in the controller.

The main example of the two-stage differential counter-pressure system is the 'Liquid' beer filler, which used to be supplied in the UK by Meyer Liquid. Like the Worssam machines, it is no longer manufactured, but spare parts are still supplied for those machines that are still in use. The construction is similar to that shown in Fig. 8.15b, except that the controller is divided into three compartments by two concentric cylinders, and the beer is carried in a separate annular supply tank. During the circuit round the filling machine, the first stage of low-pressure filling occupies

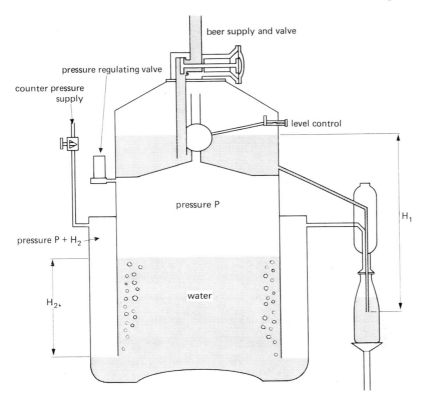

Figure 8.15b. Principle of Worssam filler

about 90° of the revolution, and the second stage of higher-pressure filling about 200°. The counter-pressure valve is opened mechanically by the movement of the bottle, and when the correct initial pressure is established in the bottle, the liquid valve opens automatically. The change-over to the second pressure occurs when the appropriate slots and holes coincide in the central manifold.

SIMPLE PRESSURE FILLERS

There are now very few pressure fillers in which the bottle is sealed against the filling head, but there are several models which pump viscous liquids under pressure without making a seal. For example, Mather & Platt's rotary pump filler, shown in Fig. 8.16, can fill 1 lb jam jars at 120 a minute. The filling assembly is in two halves. The lower half is a circular rotating plate carrying a number of nozzles, and when the nozzles coincide with an extended port in the upper (stationary) half, liquid is forced through them by a pump. Bottles or jars are led into the filler by a feed screw, and move round under the filling head in synchronism with the nozzles. The flow

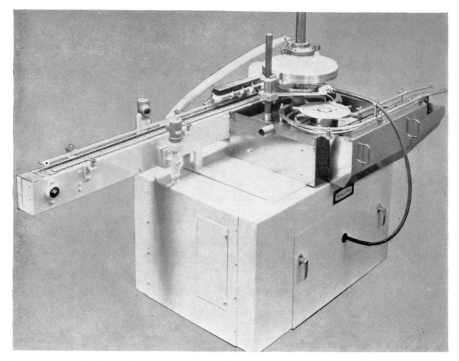

Figure 8.16.   Mather & Platt rotary pump filler

rate is determined by the pump speed, and the filling time by the conveyor speed, both of which are adjustable. If pump and conveyor run at constant speeds, the filler can be regarded as a volumetric machine, and it is in fact capable of high accuracy.

SNIFT FILLERS

The snift principle has been used for nearly a hundred years, and it evolved from the earlier 'knee bottler' for carbonated liquids, at which the operator used his knee to press the bottle up against the filling tap, and slightly released it at intervals to let some of the gas out of the bottle. A slight digression may perhaps be permitted here, to show how the famous old 'Codd' bottles, sealed with an internal glass marble, were filled. Fig. 8.17 shows the 'Hercules' single-head machine produced by W. Meadowcroft & Son Ltd., in 1891.

The bottle was placed upside-down on the wheel at '12 o'clock'. During the first half of the wheel's revolution a measured dose of up to 2 oz of syrup was pumped into the bottle. When the bottle reached the upright position, alternate filling and snifting was carried out, and by the

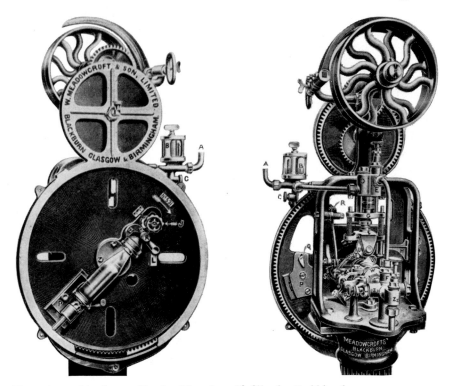

Figure 8.17.   Meadowcroft's 1891 Hercules snift filler for Codd bottles

time the bottle had returned to the inverted position, the glass marble was replaced on its seating by gravity. The bottle could then be removed, the marble being held on its seating by the internal pressure. All the valves were operated by the cam-shaft shown in the rear view. A speed of 20 bottles per minute was claimed – no wonder the machine was called Hercules! The cost of this filler was £30.

Meadowcrofts and others have developed many improved snift machines ever since the Hercules, though not many are now manufactured. In the UK the main demand comes from bottlers who want a fairly small and inexpensive machine but there used to be one large French machine (the SMA) in which *le sniftage* occurred on 54 heads at over 300 bottles per minute.

LOW-VACUUM FILLERS

Low-vacuum fillers are very widely used for milk (pasteurised and sterilised), wines, spirits and many other still liquids. A leading maker of

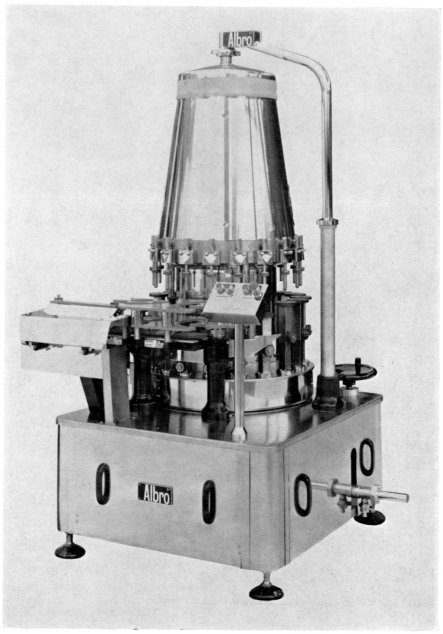

Figure 8.18.   Albro CT18 low-vacuum filler

these machines is Albro, and Fig. 8.18 shows the latest design of low-vacuum filler with 18 heads. Larger machines of this type are capable of filling as many as 600 bottles a minute.

The method of operation is exactly as shown and described in Fig. 8.9. For some liquids the filled bottle is briefly lowered away from the filling head and briefly replaced before it is finally removed; this allows atmospheric pressure into the headspace, and improves the control of filling height (measured in this case from the *top* of the bottle). Filling heights are adjusted by placing rubber washers on the filling tube.

This Albro design and other models are largely made with straight lengths of pipe, which can be opened at both ends for cleaning and for visual inspection.

ADJUSTABLE-VACUUM FILLERS

Adjustable-vacuum filling is used in several multi-head rotary machines, some capable of handling very viscous semi-liquids such as mayonnaise, hair cream, mustard and even mincemeat. The principle also lends itself well to the construction of small semi-automatic machines with one to six heads.

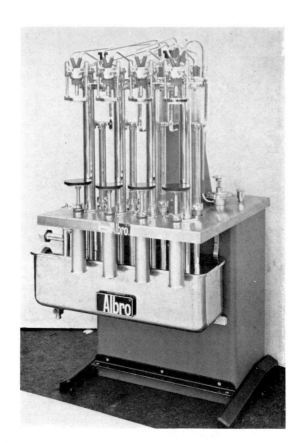

Fig. 8.19. Albro semi-automatic adjustable-vacuum filler

Fig. 8.19 shows an Albro four-head machine, capable of filling up to 40 bottles per minute. Bottles are placed by hand on the bottle lifts, and removed by hand when full, the rest of the operation being automatic. When the bottle is full, the excess liquid is drawn over to the transparent interceptor chamber seen behind each filling head, and thence through a non-return valve to the supply tank below. Before the filled bottle is lowered, the vacuum pipe is closed, and a third pipe opened to admit atmospheric air to the headspace.

FILLERS USING COMBINATIONS OF PRESSURE, GRAVITY AND VACUUM

Pneumatic Scale make a range of 'All-Purpose' fillers which can be adapted to employ any combination of pressure, gravity and vacuum. A 40-head machine is shown in Fig. 8.20. This machine uses the method, described earlier, of controlling the filling heights accurately by lowering the bottles slightly for the final suck-off. Fine adjustments to filling heights can be made while the machine is running, by means of a hand-wheel which alters the distance through which the bottles are lowered.

Figure 8.20.　Pneumatic Scale 40-head All-Purpose filler

PREDETERMINED VOLUME FILLERS

We have already discussed, under Pressure Fillers, machines which control the volume during the action of filling, but 'predetermined' volume fillers measure the liquid into a cylinder, before it is put into the bottle. The method is not widely used for non-viscous liquids in glass bottles, because it is generally slower than control by filling height, and not necessarily more accurate. For viscous liquids and semi-solid food products it is more popular; usually a piston forces the material out of the measuring cylinder. The other main application of the method is for filling very small containers, such as vials and ampoules. Fig. 8.21 shows a volumetric machine made by Autopack, for the automatic filling and capping of small pharmaceutical bottles or vials. There are two measuring cylinders and filling heads on the right of the machine, which can fill 50 vials a minute.

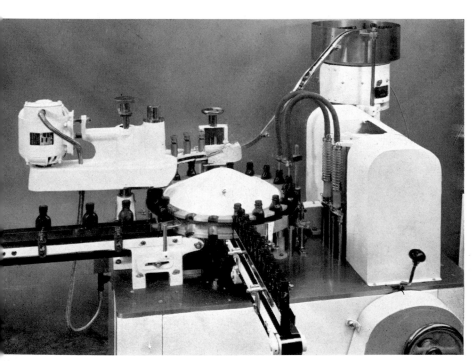

Figure 8.21. Autopack volumetric filler and capper

Filling of solids

Most dry powders can be handled by a combination of gravity and vacuum in much the same way as liquids, and a diagrammatic form of a powder filling head is shown in Fig. 8.22.

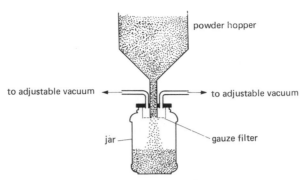

Figure 8.22.   Vacuum powder filling

The powder is held in a hopper, with an outlet pipe of dimensions such that the powder will not flow through when the air pressure is equal inside and outside the hopper. As soon as the pipe is inserted into a container which is at a lower pressure than atmospheric, flow commences. The outlet pipes are protected from clogging by a metal gauze filter, and the position of this filter determines the filling height. There may be a stage of vacuum-release or 'blow-back' to clear the gauze before the filled container is removed. The use of vacuum not only controls the flow, but also prevents the powder from blowing about. The degree of vacuum used determines the tightness of packing and hence the weight of powder obtained.

If direct control of powder weight is required, as with some pharmaceutical preparations, gravimetric fillers are available which are in effect automatic weighing machines. The accuracy of such machines has to be verified by Weights and Measures authorities, and usually has to be within 0.5%.

If a powder cannot be handled by a simple gravity-vacuum method, either because it tends to stick in the outlet pipe or because it flows too freely, then the flow can be controlled by an auger, or Archimedean screw, placed in the outlet pipe. As the auger turns it forces the material out of the pipe, working on exactly the same principle as a domestic mincing machine.

Usually the auger is preset to make a definite number of turns for each container; in this way the machine becomes in effect a volumetric one.

Fig. 8.23 shows an Arenco System 400B auger filler, while an ATD1 auger filler by Transmatic Fyllan is seen in Fig. 8.24. Single-head machines of this type provide outputs of up to 60 packs per minute. Other types of auger filler with multiple filling heads, in-line or rotary, can provide outputs of up to 300 per minute.

Figure 8.23. Arenco System 400B automatic auger filler

Variations in the bulk density of powders can cause difficulty in achieving an accurate fill with auger fillers, so the machines may need to be used in conjunction with tare weighing and check weighing systems. To avoid this, Transmatic Fyllan have developed an electronic method of weighing the powder as it passes through a convergent cone fed from an auger filler. As it falls through the cone, the powder is made to change direction, imposing a force on the cone proportional to the mass flow. The force is measured electronically and the information passed to a controller which 'memorises' the weight of powder passing through the cone. The auger is stopped when the total reaches a preset figure.

Solids in the form of separate articles such as pills or tablets are usually measured out by counting. Methods range from direct systems using a counting frame rather like a solitaire board, to indirect ones which, for example, line tablets up in a column and deliver a fixed length of column into each bottle. Also there are electronic counters, which are very accurate but not as fast as the indirect method. Alternatively tablets can be weighed out in the same way as a powder.

Figure 8.24.   Transmatic
Fyllan ATD1 automatic auger
filler

List of filling machinery

There are about 50 suppliers of filling machinery, and several hundred individual machines, consequently it is only possible to give brief details here, divided into four groups:

1. Fillers for still liquids and semi-liquids;
2. Fillers for gaseous liquids, especially carbonated beverages;
3. Fillers for solids;
4. Equipment for filling small vials and ampoules.

Information given consists of name of supplier, together with name of manufacturer if foreign; principle of operation; numbers of filling heads; maximum size of containers; maximum speed claimed (usually for one of the smaller containers in the range). Machine speeds depend on the properties of the product and the dimensions of the container, so the figures quoted may not be directly comparable. Dashes in the table indicate that it is not possible to give a figure, for example, when the speed of a semi-automatic machine depends on the skill of the operator.

Still liquids and semi-liquids

Supplier	Name of machine	Principle of operation*	No. of filling heads	Max. container capacity	Max. no. of bottles per min.
Adelphi	—	G, A-V and P-vol	1	—	—
Albro	—	L-V	4 to 72	1 gal	600
		A-V	1 to 72	1 gal	600
		V-G	4 to 72	1 gal	600
		P-vol	1 to 30	40 fl oz	70
		W	1	10 gal	30
Anglo Continental (Zanasi Nigris)	Various	P-vol	—	—	—
Arenco	LKF-1	P-vol	1	10 litres	82
	Various	P-vol	1 or 2	—	200
Autopack	—	P-vol	2	4 fl oz	50
Bancroft	Universal	A-V	1 to 6	5 litres	40
	—	P-vol	2	2½ litres	40
	Auto-Vac	L-V	1 to 24	5 litres	—
	Flo-Matic	P-vol	4 to 48	5 litres	—
Becker	Bexuda	P-vol	1 to 16	10 gal	120
Bosch (Strunck)	Various	G, V-G, L-V, A-V, vol	1 to 84	10 litres	600
			mono-block		1200

* Abbreviations: G Gravity; V-G 'Vacuum-assisted' gravity; P Pressure; L-V Low vacuum; A-V Adjustable vacuum; -vol Volumetric; W Quantity controlled by weighing.

Still liquids and semi-liquids (continued)

Supplier	Name of machine	Principle of operation*	No. of filling heads	Max. container capacity	Max. no. of bottles per min.
Bramigk (Simonazzi)	Asso Rivul	P-vol	—	1 litre	50
	Vecoce	L-V	—	—	666
	Asso	L-V	12 to 30	—	266
Cockx	Speedy	V-G and P-vol	3 to 6	40 fl oz	37
Engelmann & Buckham (Pont-a-Mousson)	Vacuum-Jet	L-V	24 to 48	—	320
	Detrez	L-V	24 to 48	—	320
Erben	Various	A-V	12 to 72	2 litres	450
Farrer	Farason	G-vol and P-vol	Various	5 litres	200
Flexile	Farmomac	L-V	2 to 18	1 litre	300
	Various	G-vol	2 to 18	1 litre	300
	Kalix KX60	G-vol	1	500 ml	35
	„ HB 60	G-vol	1	500 ml	35
	Coster 250 DS	P-vol	1	500 ml	35
FMC	Unifiller	P-vol	10 to 26	1 litre	650
Glenhove (Kugler)	VG16	L-V	16 to 18	1 litre	166
	K54	G-vol	6	550 ml	90
	Mastermeasure	G-vol	1	25 litres	—

Supplier	Name of machine	Principle of operation*	No. of filling heads	Max. container capacity	Max no. of bottles per min.
Goddard (PKB)	Biopex	P-vol	12, 16 or 32	1 litre	—
Gravfil	Various	G and P	1 to 72	5 litres	500
	Various	L-V and A-V	1 to 72	5 litres	500
	Various	P-vol	1 to 72	1 litre	500
Nealis Harrison	Rapid	A-V	1 to 4	1 gal	20
Hart (Brüser)	—	G	Spray	—	133
(Cozzoli)	Various	P-vol	—	10 litres	300
Holland (Pfaudler)	Various	P-vol and G	1 to 36	—	1200
(Simplex)		P-vol	1, 2 or 4	10 pints	100
(New Way)		P-vol or L-V	6 to 36	—	300
King	Kingmatic	P-vol	2, 4, 6, 8 and 10	1 litre	300
Mather & Platt	—	Various	1 to 32	—	600
(Rutherford)	Akra-Pak	P-vol	1	4·5 litres	110
Meadowcroft	Meadowcroft	A-V and G	10 to 32	2 litres	300
Neumo	Various	P-vol (pump)	1 to 30	5 litres	1500

* Abbreviations: G Gravity; V-G 'Vacuum-assisted' gravity; L-V Low vacuum; A-V Adjustable vacuum; P Pressure; -vol Volumetric; W Quantity controlled by weighing.

N

Still liquids and semi-liquids (continued)

Supplier	Name of machine	Principle of operation*	No. of filling heads	Max. container capacity	Max. no. of bottles per min.
Pacunion (Bausch & Ströbel)	Various	P-vol	—	500 ml	120
Purdy	Farjak	G and V-G	12 to 24	80 fl oz	180
Roberts	RF7 and 8	P-vol and W	1 to 24	1 litre	250
Rockwell	Series A, B, C	'All-purpose'	8 to 48	5 litres	480
Pneumatic Scale (Pneumatic Scale)	Pneumaflow	G and P	8 to 48	7 litres	480
	Accuflow	-vol	10 to 32	5 litres	400
Skerman	Various	—	—	—	—
Stork-Amsterdam	Stork-Votator	P-vol	4 to 48	5 litres	1200
	Various	V-G and A-V	16 to 80	—	—
Transmatic Fyllan	Various	P-vol and P-W	1 or 2	—	—
U.D. Engineering	Various	L-V	24 and 42	1 pint	110 and 200
	12/4SF	V-G	12	1 pint	40
	Various	V-G	9 to 48	1 pint	400
Vickers-Dawson	Gem-Cemac	V-G	3 and 6	1 litre	35
	Gem-Cemac	V-G	9 to 48	1 litre	440
	Gem 660	V-G	72	1 pint	660

* Abbreviations: G Gravity; V-G 'Vacuum-assisted' gravity; P Pressure; L-V Low vacuum; A-V Adjustable vacuum; -vol Volumetric; W Quantity controlled by weighing.

Gaseous liquids

Supplier	Name of machine	Principle of operation*	No. of filling heads	Max. bottle capacity	Max. no. of bottles per min.
Aerofill	Dul-Pak	P-vol	1	1 litre	15
	Star-Pak	P-vol	3	1 litre	45
	Rotary	P-vol	6, 9, or 12	1 litre	250
Bramigk (Simonazzi)	Eurostar	C-P	—	—	1666
	Rivalca	C-P	16 or 24	2 litres	133
British Syphon	Riley	S	3 and 9	Soda syphons	12
G. Clayton	Vista	C-P	18 to 42	40 fl oz	200
Flexile (Coster)	450 G	P-vol	1	450 ml	35
	6 AVG	P-vol	2	900 ml	40
	44S	P-vol	4	1800	55
	88S	P-vol	4	2 litres	100
	Rotary	P-vol	6, 9, 12	2 litres	240
Crown Cork	Crown Belcrown } Super Uniblend Super Cemco	C-P	20 to 120	1 litre	1200

* Abbreviations: S Sniffed pressure; C-P Direct counter-pressure; P Simple pressure; -vol Volumetric.

Gaseous liquids (continued)

Supplier	Name of machine	Principle of operation*	No. of filling heads	Max. bottle capacity	Max. no. of bottles per min.
Erben	Hansa	C–P	24 to 144	40 fl oz	830
Holstein & Kappert	—	C–P	20 to 164	40 fl oz	1670
	—	C–P	70	2 litres	—
Goddard (Kronseder)	Contibloc	C–P	16 to 120	1·8 litres	1370
Meadowcroft	Meadowcroft	C–P	14 to 100	40 fl oz	1000
Meyer	Dumore	C–P	24 to 84	40 fl oz	1000
Neumo	Various	P–vol (pump)	1 to 30	5 litres	1500
Stork–Amsterdam	—	C–P	Various	1 litre	1080
Vickers–Dawson	Speedwell	S	8	26 fl oz	24
	Various	C–P	24 to 100	40 fl oz	900

* Abbreviations: S Snifted pressure; C–P Direct counter–pressure; P Simple pressure; –vol Volumetric.

Solids

Supplier	Name of machine	Principle of operation*	No. of filling heads	Max. bottle capacity	Max. no. of bottles per min.
Albro	—	A-V	1 to 24	—	300
Anglo Continental (Zanasi Nigris)	CU65	C	—	—	180
Arenco	Various	Aug	1 to 4	6 lb	160
	S.500	Aug	8	—	300
	System 400	Aug	1 or 2	—	100
	PFU	W and Aug	1 to 4	—	70
Autopack	Straight-Line	Aug or W	1 to 4	—	160
	Rotor 54	Aug or W	1 or 2	—	100
Bosch	Various	Aug, C, vol	1 to 36	—	300
Driver Southall	VF3	Aug	1 or more	—	60
	SB	W	1 or more	5 kg	15 each head
	Fluidflow	W	1 or more	5 kg	20 each head
FMC	—	G-vol	—	—	300

*Abbreviations: A-V Adjustable vacuum; Aug Auger filler; C Quantity controlled by counting; P Simple pressure; -vol Volumetric; Vib Vibrating screen; W Quantity controlled by weighing.

Solids (continued)

Supplier	Name of machine	Principle of operation*	No. of filling heads	Max. bottle capacity	Max. no. of bottles per min.
Femia	Solbern	G	—	all sizes	180
Hart	—	Vib	—	—	100
(Cozzoli)	—	C	—	—	—
(Brüser)	—	G-vol	—	—	84
Holland	Various	G-vol or P-vol	1 to 24	—	300
King	—	C	Various	—	—
M. & M. Process Equipment (Herbort)	Universa 56S	Vib	—	5 kg	300
	Universa 56A	Vol	—	5 kg	300
Neumo	Various	P-vol (pump)	1 to 30	5 litres	1500
Rockwell Pneumatic Scale (Pneumatic Scale)	Vacuflow	A-V	1 to 24	—	300
Transmatic Fyllan	Various	Aug and W	1, 2 or more	—	200

*Abbeviations: A-V Adjustable vacuum; Aug Auger filler; C Quantity controlled by counting; P Simple pressure; -vol Volumetric; Vib Vibrating screen; W Quantity controlled by weighing.

Equipment for filling vials and ampoules

Supplier	Material filled	Principle*	Max. speed
Adelphi	Liquid	P-vol	16 vials/min.
Anglo Continental	Liquid	P-vol	60 vials/min.
(Zanasi Nigris)	Powder	—	416 vials/min.
(CIONI)	Liquid	P-vol	200 vials/min.
Arenco	Powder	Auger	—
	Liquid	P-vol	110 vials/min.
Autopack	Liquid	P-vol	{ 90 ampoules/min. 50 vials/min.
	Powder	Aug	50 vials/min.
Bosch	Liquid	P-vol	450 vials/min.
(Strunck)	Powder	P-vol	250 ampoules/min.
		Auger	300 vials/min.
Flexile	Liquid or powder separate or combined in same vial	P-vol	100 ampoules/min. 60–350 vials/min.
Hart (Cozzoli)	Liquids or powders	Various	120 vials/min.
King	Liquid	P-vol	100 vials/min.
Manning	Liquid	P-vol	100 vials/min.
Pacunion (Bausch & Ströbel)	Liquid	P-vol	216 ampoules/min. 120 vials/min.

*Abbreviations: Aug Auger filler; P Simple pressure; -vol Volumetric.

9. Sealing

There are literally hundreds of different ways of sealing bottles, but fortunately they have many features in common, and we can deal with them satisfactorily in groups. Without exception, closures are required to perform some or all of the following seven basic functions:

1. The bottle must be well sealed, and remain so until it is opened by the user. A 'good' seal usually means that no solid, liquid or gas shall escape from the bottle, or penetrate into it.

2. The closure should neither affect or be affected by the contents of the bottle.

3. The legitimate user must be able to get at the contents easily, normally by removing the closure but sometimes by piercing a hole through it (e.g. for blood transfusion and injection fluids). Or an orifice, spout or other dispensing device can be provided which is operated without removing the closure.

4. In some cases, would-be users must be discouraged or prevented from opening the bottle, hence the new technology of child-resistant closures for products which could be harmful to unsupervised small children.

5. The closure may have to make a good re-seal; this means that after the bottle is opened, it should be possible to replace the closure easily by hand to give effective protection to the remaining contents.

6. Some closures need to be 'pilferproof', meaning that it must not be possible to remove and replace the closure without leaving visible evidence that the bottle has been unsealed.

7. The closure should look suitable for its purpose. A well-designed and, if needs be, attractively decorated closure can considerably enhance the appearance of the whole pack.

It would be wrong to imply that the closure *by itself* has to perform these functions: in practically all cases a good seal is only obtained by co-operative action between the bottle finish and its closure. Both must be designed to fit together properly and the closure must be applied to the bottle correctly. In some applications, special design features, such as raised rings that bed into liners in the closures, are incorporated in the design of the glass finishes so as to improve the quality of seal.

It is relatively easy for a packer to choose a suitable closure if he is going to pack a product similar to one already being packed in glass by other people. He may find that there are several different caps being used successfully, and he can choose the one which suits him best. For a new product, however, a number of important technical questions have to be settled, and the packer may need the help of a specialist to work out the various possibilities and carry out the tests which will almost certainly be necessary. If a completely new pack is to be developed, the glass maker and closure supplier will need to collaborate at the earliest possible stage.

It is thus not possible to give here firm recommendations for the sealing of every product which might conceivably be packed in glass. What can be done is to discuss the properties of the available closure materials, and explain how the designer ensures that the cap will meet its requirements. In this chapter we shall be concerned mainly with the first two requirements listed above: sealing efficiency, and compatibility with contents. Finally, an outline will be given of the main types of closure available, what they are used for, and methods of application.

Types of seal

All closures fall into one of four groups, according to the conditions in which they have to operate:

1. *Normal seals.* The primary job of these closures is to provide a good seal when there is no need for the pressure inside the bottle to be different from that outside. Most normal seals will cope with the changes in internal pressure that occur with the day-to-day variations in ambient temperature – subject, of course, to the use of the correct vacuity for the type of product bottled.

2. *Vacuum seals.* These are designed to make an air-tight seal when the pressure in the bottle is significantly lower than that outside. In many cases the closure relies on the higher external atmospheric pressure to hold it firmly in position and absence of 'vacuum' can lead to leakage.

3. *Pressure seals.* These closures are designed to contain high pressures inside the bottle, of the order of 5 to 10 bars, and are mostly used for sealing pasteurised beers and carbonated beverages.

4. *Venting seals.* These are designed to contain a limited amount of pressure, but to allow some gas to escape through a vent if the pressure rises higher. When the pressure has fallen sufficiently the vent must of course close and restore the seal.

There may be some overlap between these groups; for example, pressure seals and some vacuum seals can also do the job of a normal seal. However, there is no difficulty in classifying any closure by considering its *primary* function. There are one or two closures which do not provide an airtight seal in any conditions, and are merely 'dust covers', but they are of little importance in commercial packaging and will not be discussed further.

METHODS OF OBTAINING A GOOD SEAL

The basic principles to be followed in making any demountable seal, whatever the purpose, is essentially to press a resilient component firmly against a rigid component, and to keep the resilient component in a compressed condition by continuing to apply the compressing force. It is essential that the second component should really be rigid; glass, of course, is ideal in this respect. It is equally important that the first component is sufficiently compressible and resilient. (The word resilient means that the material will try to resume its original size and shape throughout the compression, and will do so as soon as the compressing force is removed.)

In the commonest type of bottle seal a flat resilient disc or ring is pressed downwards on to the top surface of the bottle neck, hence obviously called the 'sealing surface', and the type of seal a 'top seal'. The sealing disc may be made in two parts: a sufficiently thick pad of resilient material (the 'wad'), faced with a thin layer of material which is compatible with the product. Alternatively a flowed-in liner may have been formed inside the closure; in this method the sealing medium is rendered apparently fluid by dispersing it in a suitable liquid, injecting it into the closure, and solidifying with heat.

Neither type of liner is in fact perfectly resilient, and does not need to be so. A certain amount of permanent deformation when the closure is applied can be quite helpful; this is particularly so when closures with flowed-in liners are applied hot. But there must remain sufficient resilience to keep the liner pressed firmly against the glass throughout its service.

The thickness and resilience of the liner has to be matched with the sealing force which is to be applied and with the shape and area of the sealing surface; this is a simple problem for the experienced designer.

Some of the results of solving this problem can be seen by comparing today's bottles with those made a generation ago. The old heavier bottles tended to have fairly broad and flat sealing surfaces, so their closures needed rather thick and soft liners. Today it is possible to make finishes with quite narrow and smoothly rounded surfaces, and it has been possible to reduce considerably the liner thickness and change to rather less resilient sealing materials.

Apart from top seals, it is possible to wrap the liner round the finish to some extent, producing a 'top-and-side' seal. Some types of vacuum closure rely solely on an annular compound pressing inwards on the side of the finish, thus making a 'side-seal', and, of course, there can be a resilient member inside the bottle neck pressing outwards, as with the corked wine bottle.

The methods used to apply and maintain the compressive force also vary considerably, as shown in the following examples:

In a pre-formed screw cap the liner is compressed by the screwing operation. The compression is maintained because the cap is prevented by friction from unscrewing.

With a 'rolled-on'* screw cap, the compression is obtained solely by applying top force during the capping operation; it is maintained by forming retaining threads on the side of the cap.

With 'crimped-on' closures, such as the familiar crown, the compression is again obtained with external top force, then maintained by bending the metal skirt inwards to lock it in position. Vacuum closures, too, are pressed into position with external top force, then they rely partly or entirely on the higher external atmospheric pressure to maintain compression.

With traditional cork plugs, compression is ensured by making the cork larger than the orifice into which it is forced. Sometimes the cork is tapered, so that the compression increases progressively as the cork is inserted.

COMBINATION CLOSURES

In certain applications, double closure systems are used. For example, metal or plastic capsules or shrinkable bands are applied to the necks of most wine bottles, not only for decoration but also to improve the security of seal and to provide a pilferproof feature.

The use of a membrane stuck to the mouth of the container is also a combination closure in that when the membrane has been torn from the

* The principle of the rolled-on cap is explained on page 215.

rim of the container, the cap itself provides the re-seal. (The membrane is in fact the only type of seal which does not rely on resilience and pressure.)

RESILIENT SEALING MATERIALS

Cork.   Cork has been used for over 300 years as a resilient bottle seal and is still extensively employed for sealing wine and champagne bottles. The world's whole supply of cork comes from Portugal, Spain and a few neighbouring Mediterranean countries. Cork contains many tiny air cells – of the order of 20 million per cm³. When it is compressed, the cells are flattened but not burst. As soon as the force is removed, the air pressure in the cells reasserts itself, giving a high measure of resilience. In addition to the use of cork as solid plugs, 'composition cork' has been a very successful material for making resilient wads; this consists of granules of cork bonded together into blocks with either a gelatine glue or a synthetic resin and then sliced into thin sheets.

With the development of glass finishes with narrower and smoother sealing surfaces, it has been possible to replace 'compo-cork' to a large extent with pulpboard, which is thinner, less resilient, and of course cheaper.

Pulpboard.   Pulboard is essentially thick paper and is available in several qualities, both compact and bulky, and in a range of thicknesses. Its main constituent is woodpulp. Pulpboard, either plain or laminated to facing materials, is now the most widely used sealing liner.

Rubber.   Natural and synthetic rubbers can be obtained with almost any degree of resilience or rigidity that may be required. Natural rubber suffers from the disadvantage of having a very high coefficient of friction and causes caps to spring back after tightening unless the sealing surfaces are lubricated with water before applying the closures. Rubber, however, has the advantage that it will withstand processing at high temperature and is self-sealing after piercing with a hypodermic needle. It has always enjoyed extensive use in pathological and blood transfusion closures, but its increasingly high cost is encouraging users to seek alternative materials.

Plastics.   Plastic sealing materials vary widely in their properties, from the fairly hard polyethylenes to the softer ethyl vinyl acetates (EVA), ionomer resins (Surlyns) and the wide range of resilience obtained with different degrees of plasticisation of polyvinyl chloride (PVC). Expanded plastics further increase the range of suitability; these are materials in which very many small bubbles have been introduced while the material is molten.

Plastics such as polyethylene can also be moulded into hollow shapes which are deformable and therefore have greater resilience than the solid material.

Plastisols and water-based compounds (Flowed-in compounds). Plastisols are dispersions of PVC resins in plasticisers. They are injected into metal caps either in the form of annular sealing rings or as overall lining discs. As discussed earlier, they are applied in liquid form and solidified by heating; this process leaves them firmly attached inside the closures.

Water-based compounds are colloidal dispersions of either natural or synthetic rubber in water. They are injected into metal caps in exactly the same manner as the plastisols, but need longer heating to drive off the water and leave solid and homogeneous sealing liners in the closures.

Flowed-in liners are almost universally used in vacuum closures as well as for all heat-processing applications. They are also increasingly replacing conventional liners.

COMPATIBILITY OF CLOSURE AND CONTENTS

None of the materials in the closure that may come in contact with the product packed must in any way taint it or affect it physically or chemically. Nor should the product itself have any effect on the materials in the closures.

Some resilient materials, such as pulpboard and composition cork, are not sufficiently resistant to aqueous products by themselves, but they can be made satisfactory by lining with a suitably resistant facing material. Most other resilient materials can be used in direct contact with aqueous products.

FACING MATERIALS

For the majority of uses composition cork and pulpboard wad materials have to be lined, as explained above, with a facing material that is compatible with the product packed and which prevents the product from contacting the wad. Facing materials must also bed down well on the sealing surface without creasing, so as not to upset the seal.

Almost all facing materials are based on paper, such as bleached sulphite or bleached kraft, coated with white-pigmented synthetic resins such as the vinyl copolymer (PVA/PVC) or the polyvinylidene chloride (PVDC) resins; or the paper can be laminated to plastic films such as PVDC (e.g. Saran), polyester (Melinex), and polyethylene. Metal foils (aluminium and tin) make excellent facing materials where maximum impermeability to gases and moisture vapour or resistance to solvents are required; they are usually backed with paper to assist their lamination to the wad material during cap manufacture. Where there is danger of the product corroding the metal, the foil can be lacquered on its product-facing side or laminated to a polyester or PVDC film.

The range of facing materials used to be more extensive than it is today, but an endeavour has been made to cut down the number in the interest of rationalisation. Closure manufacturers have always sought to find a universal facing material or liner for all products but this ideal is not yet in sight.

RING TYPE SEALS

If the resilient component is a complete disc, no other part of the closure comes into contact with the contents, but if it is in the form of a ring, the behaviour of the inside of the top of the cap must also be considered. This applies particularly to metal vacuum closures for wide-mouth jars. The inside lacquer or enamel system with which the cap is protected must therefore be compatible with the product packed not only at room temperature but also under the heat-processing treatment to which the pack itself is likely to be submitted. The same question may arise with plastics closures, which are sometimes fitted with rubber sealing rings.

Closure Design

A closure is designed for its appearance and performance. In this context, 'closure design' includes the choice and testing of materials for the external and internal parts, planning the dimensions and tolerances, and deciding on the appearance and decoration required.

CLOSURE MATERIALS

Metal. The basic metals for closure manufacture are tinplated and tin-free steel, pure aluminium, and some aluminium alloys. All metal caps are stamped and drawn out of sheet or coil which has first been lacquered or enamelled on the inside surface for protection, and also enamelled, printed and varnished on the outside surface for decoration. The sheet can also be embossed during the stamping operation to provide other decorative effects.

All metal caps have raw edges where they are stamped out of the sheet. These raw edges are usually curled away out of sight but liquids and vapours can still penetrate to them. Consequently, steel-based caps should be kept away from rusting environments, and if the caps are wet after processing, they should be allowed to dry completely before the packs are stored. Aluminium does not rust and is resistant to acid conditions but it is sensitive to alkaline environments.

Crown closures are almost always made in tinplate because of its high strength and rigidity; a large proportion of prethreaded screw caps and the great majority of lug caps are also made in this material.

Due to its softness, aluminium and its alloys are ideal materials for roll-on and spin-on caps. It is also particularly suitable for deep drawn caps such as the Stelcap which combines the functions of a rolled-on closure and a capsule. Attractive decorative effects are possible on the skirts of Stelcaps by side printing, embossing and shaving (i.e. removing the enamel from raised embossing, leaving a bright metallic surface). Aluminium and its alloys can be used for making prethreaded screw caps, and aluminium is also used for crimped-on vacuum closures.

Thermosetting plastics. These have been used for many years for a very wide range of moulded closures. 'Thermosetting' describes a material which sets permanently into a hard mass when heated under pressure. The usual moulding powders consist basically of either urea–formaldehyde or phenol–formaldehyde; they are formed into caps by compression moulding, i.e. by applying heat and pressure in a mould. Any colour required is incorporated in the moulding powder, and there is considerable scope for good design and decorative effects.

The ordinary urea plastics tend to absorb a little moisture and to lose it again in dry conditions, which makes them swell or shrink slightly. They are, therefore, not recommended for use in climatic extremes, nor for processes involving wet heat treatment. The phenolic plastics are more resistant in this respect; their only limitations are that the range of colours possible is restricted to dark shades, and some phenolic materials may impart a flavour or odour.

Thermoplastic materials. These are used to make injection-moulded closures (the materials melt when heated and are forced into the mould under high pressure). The flexible kinds, like polyethylene, are mainly suitable for snap-on closures, while for screw caps a more rigid material, like polystyrene, is desirable. The chemical and dimensional stability of thermoplastic materials is generally better than that of the thermosetting ones, but the resistance to heat is, of course, lower and the solvent resistance is not as good.

SELECTION AND TESTING OF CLOSURES

When selecting closure materials it is obviously necessary to have full information about the product to be packed and the container to be used; also the extreme conditions of atmosphere and temperature they may encounter, either in processing or in service.

It is possible to carry out accelerated tests for some purposes; for example, to determine at what temperatures, pressures and vacuities the filled and capped bottles are liable to leak. It is also possible to reproduce the processing conditions carried out by the customer; but for reliable assessment of compatibility of closure and contents there is no real substitute for long-term storage tests, using exactly the same type of closure, bottle and contents as will be used in practice. Closure testing laboratories should be equipped to do both accelerated and long-term tests, and to reproduce all conditions of temperature and humidity which caps may encounter anywhere in the world.

Some bottle finishes are more difficult to make than others, and this will be reflected in the price; so when choosing between closures on a cost basis the packer should check on the corresponding bottle costs. Three features will tend to increase bottle costs: a very complex surface, for example with deep threads close together; a sharp under-cut or re-entrant in the glass surface; and a design which requires a greater thickness of glass in the finish than in the rest of the neck and body.

DIMENSIONS AND TOLERANCES

A closure will only perform correctly if it is in the right position on the bottle, and if its resilient component is compressed correctly. This immediately brings us to the question of dimensions and tolerances for bottle finishes and closures.

A great deal of work has been done on bottle finishes, and toleranced specifications have been agreed for every important type. Two series of glass finishes have become the subject of British Standards:

BS 1918: Part 1, Continuous-thread Screw Finishes
BS 1918: Part 2, Crown Finish

In addition numerous specifications for other standard finishes have been drawn up by the Glass Manufacturers' Federation in consultation with the closure manufacturers. This work was started in order to assist transition to the metric system, and to date the GMF have published 45 detailed specifications for most of the important types of finish. The drawings have been classified as follows:

> 100 series – Non-screw finishes
> 200 series – Multi-start screw finishes
> 300 series – Single-start screw finishes
> 800 series – Miscellaneous finishes
> 900 series – Supplementary information

It might be expected that the tolerances for closures to fit the standard bottles would have been agreed and published in the same way, but surprisingly not much progress has yet been seen. The British Standards referred to above do not mention the closures, except in one place to say that they must fit the bottles. This may have led some people to think that cap tolerances are unimportant, but this is far from the truth: it is essential to know how much variation there will be in both bottle and closure when calculating what dimensions they should each have, in order to give the best overall performance. Appendix D explains in detail how this calculation should be made. In most cases closure manufacturers aim for a clearance of around 0·1 mm between the maximum finish and the minimum closure dimensions.

Closure application methods

We have already shown that closures fall into four main groups according to their primary functions; they can also be classified according to their method of application, namely:

Method of application	*Types of closure*
1. Screw-on, and screw-in.	Internal and external screw caps and lug caps.
2. Push-in, and press-on.	Driven corks and flanged stoppers: plugs; snap-on caps and other forms of press-on friction fit caps.
3. Crimp-on.	All closures which are secured to the rims of bottles by squeezing the skirt of the cap into a groove or under a bead on the side of the neck of the bottle.

Method of application	*Types of closure*
4. Roll-on, and spin-on.	All caps in which the screw threads are formed in the skirts by means of spinning rollers that shape the aluminium skirts to match the glass threads. Also all threadless closures in which the skirts are rolled under retaining rings on the glass necks.

The three main types of closure (normal, vacuum and pressure) can use all the above methods of application, but roll-on vacuum closures are not now common. The other eleven combinations are discussed below, together with a section on venting closures.

NORMAL SEAL: METHOD I (SCREW-ON)

Everyone is familiar with screw caps, and their wide range of uses is continually expanding.

Preformed screw caps can, if necessary, be put on and tightened by hand, but for consistent results some form of mechanical cap tightening is always to be preferred. The simplest apparatus consists of a single revolving head; a loosely capped bottle is held by hand under the head which descends, grips the cap and tightens it. The semi-automatic stage of capping is represented by a single-head machine, fed by a conveyor and simple starwheel, with the caps being loosely applied by hand as before. In automatic machines, single-head and multi-head, the caps are fed automatically from a hopper on to each bottle neck as required.

Speeds depend on the type of machine and cap; as a general guide 30 caps per minute is reasonable for a hand-fed single-head machine, rising to about 60 if the caps and bottles are fed in automatically. Multi-head machines can reach 240 a minute or more – for example, the four-head Dicomulti capper supplied by Dico Packaging Engineers, shown in Fig. 9.1. An interesting feature of this machine is the combination of worm feed and electronic proximity switch to time the arrival of bottles.

In addition to the method of tightening by means of a rotating head on top of the cap, it is possible to turn the cap with moving rollers or bands applied to the sides of the cap. If this can be done to bottles while they are moving rapidly in a straight line, it makes very high-speed operation possible – in the region of 400 caps per minute. The principle is also used at slower speeds; for example, in the Autopack filling and capping machine shown in Fig. 8.21.

To obtain consistent cap-tightness, it is necessary to control the amount

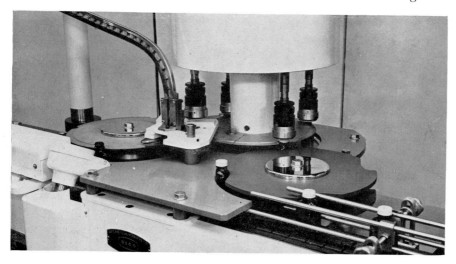

Figure 9.1.   Dico four-head screw capper

of torque* applied, and this is usually done by providing an adjustable friction clutch in the capping head, which slips when the turning force on the cap reaches the required level. For best results, it is also desirable that the downward force on the cap does not vary, because this would affect the amount of friction between the cap threads and the glass, and could prevent the cap from reaching the intended position. (If the top pressure is applied through the friction clutch, any variation in pressure will also alter the amount of torque at which the clutch will slip.)

During capping the bottle should be gripped firmly to prevent it from turning under the cap tightening head. The cap itself must also be held firmly in the capping head during the cap tightening operation; if the slip occurs at either of these points instead of at the friction clutch, all control will be lost.

When setting up a capping machine, a method is needed of adjusting the application torque on each capping head. Unfortunately, it is not practicable to measure the torque directly while the machine is working, and an indirect method has to be used.

When caps are tightened or loosened by hand, the torque can be measured, and it has been found that there is, for any type of cap, a reproducible relation between the torque required for tightening and that required for loosening. For example, if satisfactory tightness can be obtained with 25 pound-inches, all the caps so tightened might be removable with, say, 15 pound-inches.

A convenient instrument for torque measurement is shown in Fig. 9.2.

* A definition of torque is given in the Glossary.

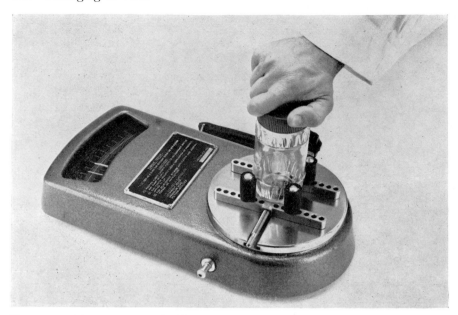

Figure 9.2.   Torque tester, U.G. Closures & Plastics

The application torque required depends mainly on the cap diameter, and to a lesser extent on the type of liner, strength of cap, pitch of threads, etc. This does not, however, mean that extreme precision is needed; in practice the torque can vary over a small range without harm. As a general guide, the following figures have been found satisfactory.

Closure diameter (mm)	Application Torque (pound-inches)	(newton-metres)
15	6– 9	0·7–1·0
20	8–12	0·9–1·4
24	9–15	1·0–1·7
28	10–18	1·1–2·0
31	11–20	1·2–2·3
33	12–21	1·4–2·4
38	15–25	1·7–2·8
43	17–27	1·9–3·1
48	19–30	2·1–3·4
53	21–36	2·4–4·1
58	23–40	2·6–4·5
63	25–43	2·8–4·9
70	28–50	3·2–5·6
89	36–64	4·5–8·0
120	48–86	5·4–9·7

Screw caps are usually made with a single continuous thread which should be of sufficient length to ensure that at least a full turn of thread engagement is obtained, particularly where tightness of seal is critical. This will ensure uniform pressure around the circumference of the resilient liner (see Fig. 9.3). Moulded caps with multi-start threads allow for a faster lift-off and quicker unscrewing and tend, therefore, to be used for the less critical seals.

Screw caps with membrane seals are supplied with membranes temporarily attached to the inside of the waxed pulpboard wad. Some membranes are sealed to the jars by glue, others by a high-frequency induction heating method known as 'Lectraseal'.

Glued membranes are made from a special paper laminate, sometimes combined with aluminium foil. Before capping, the glass sealing surface is given a thin film of glue by means of a roller; the rest of the capping operation proceeds in the same way as with any other sort of screw cap. As the glue dries out, the membrane adheres to the mouth of the jar, and separates from the waxed pulpboard wad when the cap is finally unscrewed.

The more modern 'Lectraseal' membrane consists of an aluminium foil disc that is coated with a thermoplastic adhesive and is then lightly attached to a waxed pulpboard wad in a plastics screw closure. Caps fitted with 'Lectraseal' membranes are applied to the jars in a conventional way and the capped packs are then passed under a coil connected to a high-frequency generator. The heat developed in the membranes by the induced eddy currents causes the thermoplastic to soften and the membranes to adhere to the sealing surfaces. 'Lectraseal' membranes provide an excellent hermetic seal, and the clean peel obtainable is an important advantage compared with the condition of sealing surface left after removal of a glued membrane. The process cannot be used with metal caps, because the metal upsets the high-frequency heating of the membrane.

NORMAL SEAL: METHOD 2 (PUSH-IN)

Little need be said about plain corks, which have been used for wine bottles for a very long time and are still preferred for sealing quality wines, even though other closures are being increasingly used for some of the cheaper wines. Cork is not porous in the ordinary sense of the word, but, after long storage of a filled bottle on its side, the cork becomes saturated and a little liquid seeps through. For this reason, and for decorative and protective purposes, a capsule may be used to cover the cork.

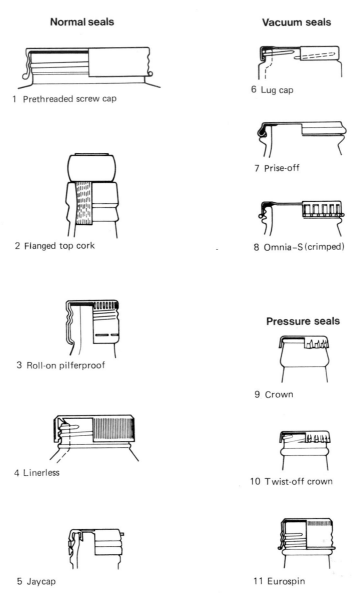

Normal seals

1 Prethreaded screw cap

2 Flanged top cork

3 Roll-on pilferproof

4 Linerless

5 Jaycap

Vacuum seals

6 Lug cap

7 Prise-off

8 Omnia–S (crimped)

Pressure seals

9 Crown

10 Twist-off crown

11 Eurospin

Figure 9.3.   Normal seals, vacuum seals and pressure seals

Because of the tightness of fit of the driven cork, the strength of the ordinary wine bottle, and the slight porosity of the cork, the wine bottler does not need to worry about vacuity. When short-flanged corks, frequently called 'flanged stoppers' (Fig. 9.3), are used, this is no longer true, since an internal pressure of 0·5 bar or less may well be sufficient to blow the cork out. Nevertheless flanged corks are still popular with packers and users, and with a modern plastic top of, say, polystyrene, they can look very attractive.

Mechanical corking machines are known curiously as 'floggers'. Among the most popular examples of these is the range made by Bertolaso. Throughputs depend on the kind of bottles being handled, but can range from 17 to 50 per minute on an automatic single-head machine and from 60 to 367 per minute on a rotary 12-head unit. The usual method of operation is for the corks to be fed from a hopper and cleaned by a jet of air before being compressed by four stainless-steel jaws and automatically inserted into the neck of the bottle.

NORMAL SEAL: METHOD 3 (CRIMP-ON)

Crimp-on closures are used more for vacuum and pressure seals than for normal seals although there is basically no reason why a crimp-on closure should not be used for a normal seal when an application arises.

NORMAL SEAL: METHOD 4 (ROLL-ON OR SPIN-ON)

The leading manufacturers of roll-on closures supply both the caps and the capping machinery required. The roll-on closure has to be made from a thin and soft material which can easily be shaped, and aluminium or one of its alloys is the most suitable.

When spun on to a bottle finish, the ordinary roll-on cap is the exact equivalent of its corresponding preformed aluminium screw cap both in performance and appearance. There are also very widely used pilferproof versions; these have security rings that break away partially or completely from the caps as soon as the closures are unscrewed.

The caps are supplied as unthreaded shells which have been fully decorated and fitted with liners. During application the liner is compressed by the top pressure exerted by the capping machine head while spinning rollers rotating round the skirt of the cap press the threads in the side wall, using the glass thread on the bottle as a former. A range of capping machines is available suitable for all line speeds – one of the fastest being a 12-head fully automatic version capable of applying caps at up to 600 per minute.

VACUUM SEAL: METHOD I (SCREW-ON; TWIST-OFF)

When unscrewing a normal screw closure a lateral frictional force has first to be overcome before the lift provided by the thread of the bottle raises the liner off the glass finish. An ordinary screw closure fitted with a suitable liner might make a good vacuum seal, but, in practice, the closure would be difficult to remove because the internal vacuum would increase the frictional forces holding the liner on the finish of the bottle.

The 'twist-off' vacuum seal, used widely throughout the world for food packaging, overcomes this problem by making use of a four-start thread of coarse-pitch on the finish of the jar. The closure itself – called a lug cap – is lined with a flowed-in plastisol sealing ring. The rapid lift provided by the four-start thread in conjunction with the low friction of the sealing compound enables the vacuum to be released without undue effort when the cap is twisted off.

These vacuum closures have found very wide acceptance, particularly for those foods not necessarily consumed all at once, such as pickles, chutneys, sauces, salad creams, jams and spreads. The caps are usually made of tinplate to provide the rigidity required by the lugs to ensure a good seal, but for some applications aluminium alloy caps could be satisfactory. They are decorated in exactly the same manner as other metal caps, but as the cap has a smooth external skirt, it presents a very smart appearance which has undoubtedly contributed to its success.

Lug caps are usually applied by the 'vapour-vacuum' method. Steam is used to pre-heat the cap and slightly soften the sealing ring immediately before the cap is applied to the jar. Steam is also injected in and around the neck of the jar as the cap is being applied, thus replacing the air in the headspace of the jar with steam; this develops a vacuum in the container when the steam condenses.

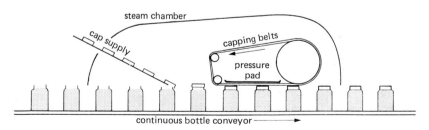

Figure 9.4.   Principle of Metal Box straight-line capper

The 'vapour-vacuum' capping machine supplied by Metal Box is of unusual design (Fig. 9.4). The caps are pressed down into position by a moving belt while the jars travel continuously along a conveyor; the twisting action is obtained by dividing the belt in two along its length and running one half faster than the other. The speed of the machine is determined mainly by the rate at which caps can be supplied from the hopper; 200 to 250 per minute is quite normal, but much greater speeds are possible if a sufficiently large cap hopper is used, or if the hopper can be automatically replenished as it is emptied.

VACUUM SEAL: METHOD 2 (PUSH-ON)

One of the simplest and oldest push-on vacuum closures is the 'prise-off' cap illustrated in Fig. 9.3. The prise-off cap is side-sealing and makes use of a butyl-rubber gasket crimped into the inside of the skirt. It is applied to the appropriate glass finish by downward pressure only and is retained by the vacuum developed in the pack by steam flushing as it is applied. The machine illustrated in Fig. 9.4 is commonly used for this purpose, but without the split belt. The cap requires a coin or similar tool to remove it and is not really intended to serve as a means for resealing a container. It is still a worthy competitor of the twist-off vacuum cap – especially when the packer wants to discourage the consumer from resealing part of the contents.

Another closure recently developed is the PT (Press-on, Twist-off) cap; this was patented by Continental Can, and is also supplied in the USA by Owens-Illinois under the name 'Vapak 3'. It is made in the UK by Metal Box.

The PT cap is a smooth-skirted tinplate cap lined with a plastisol compound placed so as to form a ring inside the top of the cap, covering both the internal radius and the inside of the skirt. It is applied to containers with a multi-start thread finish by the vapour-vacuum process. During the heat-processing of the pack the plastisol softens and takes up the profile of the glass threads. The rest of the plastisol makes a top and side seal on the finish of the container. On cooling the negative thread pattern is 'frozen' in the plastisol and the cap is removed by a normal anti-clockwise twist-off action.

Lastly, there are a number of other shallow-skirted push-on vacuum closures, lined with plastisol sealing rings, which are used for sealing jam jars and tumblers.

VACUUM SEAL: METHOD 3 (CRIMP-ON)

There are several versions, all with flowed-in plastisol sealing rings which seal on top of the jar. They are mainly used in the UK for jams and preserves, but with an appropriate lining material, they are also suitable for most foods processed in the jar.

The Omnia cap, by Thomas Hunter, is usually made of aluminium to fit jam jars and other similar finishes. The capping and crimping head makes a number of indentations in the edge of the cap's skirt, which hold the cap firmly in position. When the vacuum is released the cap can be pushed off by hand. If a good reseal is required the Omnia S cap is available. This has a deeper skirt and is crimped on to a screw-threaded glass finish. The crimps in effect make the closure into a kind of lug cap which can be unscrewed and replaced a number of times before the lugs become flattened (Fig. 9.3).

A crimped-on vacuum seal can in some circumstances act as a venting system; when the internal pressure exceeds the external pressure, the closure will lift slightly, allowing air or vapour to escape, and it will then reseal as the external pressure exceeds the internal pressure on cooling to room temperature. This feature can be used to advantage in retort processing, as discussed in Chapter 10.

PRESSURE SEAL: METHOD 1 (SCREW-ON AND SCREW-IN)

The internal screw stopper is the oldest member of this group; its rubber washer made a very good seal, and it resealed just as effectively when replaced by hand. It is the only example considered so far of a multi-trip closure. However it has now been largely superseded in the carbonated beverage trade, mainly by roll-on metal closures with flowed-in plastisol liners.

Pre-threaded metal or moulded caps have also had a period of popularity for sealing carbonated beverages (the moulded ones were returnable) but they, too, have now largely been superseded.

PRESSURE SEAL: METHOD 2 (PUSH-IN AND PRESS-ON)

Successful attempts have been made to design a plastics snap-on cap which will withstand internal pressure, but these have not been used very extensively. Caps are, however, available for resealing purposes after the crown cork has been removed.

The champagne cork, which is wired into the mouth of the bottle, is probably the oldest pressure seal used commercially and its use for sparkling wines continues today, although it is being partially replaced by finned or ringed polyethylene stoppers which also have to be wired into the mouths of the bottles. There is no such thing as a push-in seal that will contain pressure without external support (this includes the old porcelain 'swing-stopper' which is still used in parts of Europe).

PRESSURE SEAL: METHOD 3 (CRIMP-ON)

The two words 'crown cork' are nearly all that needs to go in this section.

It would be impossible to estimate how many crowns have been used throughout the world since this closure was invented by William Painter in 1892 or even how many will be used in the future.

The modern conventional crown is generally made of tinplate and is now invariably lined with a flowed-in plastisol liner that covers the inside head of the cap or in in the form of an annular sealing ring.

A recent variant on the ordinary crown is the 'twist-off' crown. This is applied in much the same way as an ordinary one, but the glass finish is threaded in such a way as to make it possible to twist the crown off by hand. (It is not intended to provide a proper reseal action.) See Fig. 9.3.

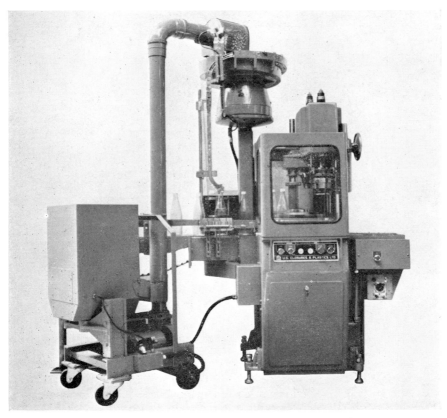

Figure 9.5.   Twin-head roll-on capper (Method 4), U.G. Closures & Plastics

PRESSURE SEAL: METHOD 4 (ROLL-ON)

Several types of roll-on closure, such as Flavorlok, Eurospin, Coronet, etc., have now become the most widely used group for carbonated soft drinks (Fig. 9.3). In some countries pilferproof versions of these are much used.

The construction of the cap and liner and the method of pressing the liner into position on the bottle have to be carefully controlled to obtain maximum pressure retention. A number of different automatic capping machines with from one to eight capping heads are available. Fig. 9.5 shows a twin-head U.G. machine, which provides throughputs up to 100 bottles per minute.

VENTING SEAL: METHOD 1 (SCREW-ON)

Developed originally for sealing bleaches (hypochlorites) and peroxides, venting seals are designed to relieve excess pressures in bottles. A venting closure is basically a conventional external screw closure fitted with a liner that vents. The first venting liners consisted of rubber discs with central slits, which under the effect of internal pressure arched into a dome in the top of the closure; the slit opened and the excess gas pressure escaped. Next followed discs in plastics and similar materials provided with grooves at the back that crossed the sealing edges of the bottle finishes. As soon as excess pressure developed in the bottle the grooves arched and the excess gas pressure escaped.

There are numerous variations on the groove method of venting, and many other venting devices are covered in the patent literature. One that shows promise is a porous plastics liner that allows gases to escape via the pores when excess gas pressure develops.

Bottles with venting seals need to be kept upright always, to avoid the risk of liquid leakage when the vents open.

Special-purpose closures

The many kinds of special-purpose closure now in use come within the groups and sub-groups discussed above, but one or two of the more important systems need to be discussed separately in more detail.

LINERLESS CLOSURES

Linerless closures have always been attractive in terms of closure economy, but there is a conflict of requirements between the rigidity needed in the body of the closure to enable it to be screwed on tightly and the resilience required of the sealing element. Polypropylene is one of the few materials that possess the degree of balance between rigidity and resilience that makes linerless closures feasible. It also has excellent chemical resistance and is suitable for use on a wide range of food, pharmaceutical, cosmetic and household products.

One of the most common types of linerless closure consists of a conventional screw closure with one or more moulded annular rings, fins or claws just above the sealing surface of the bottle (Fig. 9.3). These flex resiliently against the sealing surface when the cap is tightened.

Linerless closures have also been designed to make a bore or plug type seal in the finishes of glass bottles. A cap developed in France has a double internal skirt consisting of a thin flexible skirt backed by a thicker more rigid one. When the cap is applied to the bottle, the flexible skirt presses resiliently against the wall of the bore and the more rigid skirt provides the back-up necessary to maintain the sealing pressure. A bore type seal can also be combined with a top seal.

Another example of a linerless closure is the snap-on cap. This is designed to snap over a corresponding bead provided on the finish of the bottle or jar. Snap-on closures are generally made in polyethylene, and the tightness of their fit determines the effectiveness of the seal.

The Jaycap is a popular snap-on polyethylene closure with special features that have made it very attractive in the marketplace. It is supplied by Johnsen & Jorgensen and has a triple sealing feature: a plug seal inside the bore; a top seal on the rim; and an outside seal under the snap-over bead (Fig. 9.3). Between each sealing point there is a capillary cavity that prevents seepage should the product penetrate the initial plug seal. Jaycaps are applied by simple top pressure; they are provided with a pilferproof band which has to be torn away before the cap can be opened. The movable cap is connected to the fixed portion of the closure by a hinge, and the cap itself is held open by a small lug while the contents are being poured out. It has found wide usage in the soft drinks, pharmaceutical, cosmetic and food industries.

CHILD-RESISTANT CLOSURES

A child-resistant closure should have an opening action sufficiently complex to defeat a small child's efforts, but an adult should not have any difficulty with it. Standard tests for closure security have been specified in BS 5321. At least five different operating systems have been used so far:

1. *Press-and-turn.* The cap must be pressed against the bottle and rotated at the same time.

2. *Squeeze-and-turn.* A double cap with a free rotating outer cap which must be squeezed to engage the inner one when the closure is to be unscrewed.

3. *Combination lock.* Two parts of the cap must be lined up, according to the marks shown on the cap.

4. *Restraining ring.* A two-piece cap in which both parts (top and restraining ring), when screwed together, rotate freely on the bottle. As the restraining ring is permanently attached to the bottle, the latter can only be opened by holding the restraining ring in one hand and un-screwing the top of it with the other hand.

5. *Press-and-lift.* Downward pressure relaxes the grip of the cap's edge on the bottle.

Some child-resistant systems require special bottle finishes, but an important closure which can be used on standard glass finishes is the 'Clic-Loc' made in Britain by U.G. Closures & Plastics. This is a 'press-and-turn' closure – two co-ordinated movements that almost always baffle a child under five. The Clic-Loc has the further advantage that it makes a loud clicking sound if a child attempts to unscrew it by a normal turning action, thereby alerting the parent. Clic-Loc closures are now being used on an increasing number of potentially dangerous medicines and toxic household products.

WIDE-MOUTH BEVERAGE CLOSURES

There is a consistent demand for easy methods of opening beverage con-tainers and several methods have been developed for wide-mouth glass containers by which the aluminium closure is crimped or spun on, and then removed by a tearing action. A ring or tab is needed to start the tear, which should continue along scored lines. Two examples in UK use are the Rip Cap, supplied by American Flange, and the Maxi Cap, supplied by Metal Closures.

A more novel system is provided by the Seidel Strip closure developed by United Glass; with this the cap is held on after spinning by a security band round the skirt, and it can easily be released by peeling off the band. Examples of packs with Seidel Strip closures are shown in Fig. 16.3.

Sealing Machinery List

This list does not include machinery for applying crowns, or aluminium caps on milk bottles, as this equipment is normally an integral part of the filling machinery. Machines for inserting corks are included only if they have been designed solely for this purpose.

Normal and pressure seals

Supplier	Machine name	Type of closure	Details of machine*	Max. no. of seals per min.
Albro	—	Screw or press-on	1, 4 or 10 heads, A	60, 120 or 300
	—	Snap-on	1 head, A	80
	—	Snap-on	Overbelt, A	200
	—	Crown	Monobloc, A	300+
	—	Snap-on	Monobloc, A	200
	—	Roll-on	Monobloc, A	300
	—	Cork stoppers	1 head, S/A	60
	—	Cork stoppers	Overbelt, A	100
Anglo	TU63	Plug	1 to 4, A	180
Continental	AU69	Screw	1 to 6, A	180
(Zanasi Nigris)	GU63	Crimped	1 to 6, A	180
Arenco	—	Screw	1, A	35
Autopack	—	Screw Plug or Snap-on	S/L	130
Bancroft	—	Screw	1 head, S/A	32
Bosch (Strunck)	Various	Plugs, roll-on, screw and press-on	1 to 18 heads, S/A and A	600
		snap-on, press-in rubber bungs	Monobloc	1200

* Abbreviations: S/A Semi-automatic; A Automatic; H Hand capper; S/L Straight line (as opposed to rotary head).

Normal and pressure seals (continued)

Supplier	Machine name	Type of closure	Details of machine*	Max. no. of seals per min.
Dico	Dicoroll	Roll-on	1 head, A	60
	Dico-multiroll	Roll-on	4, 6 or 8 heads, A	270
	Universal	Screw	1 head, A	60
	Dicomulti	Screw	4, 6 or 8 heads, A	300
	Dicoseal	'Clinched' seal	1 head, A	240
	Single head	Snap-on	1 head, A	150
	Dicomultihead	Snap-on	4, 6, 8 or 10 heads, A	400
Engelmann & Buckham (Pont-a-Mousson)	Detrez	Crimped and screw	A	—
Fairest Trading (Calumatic)	Various	Screw, roll-on, press-on	1 head, A 3 heads, A	50 125
Flexile (Farmomac)	F15 Various	Screw Screw, push-on, roll-on, plug	1 head, S/A 1 to 8 heads, A	— 300
Glenhove (Datz)	SVM	Roll-on, or screw	1 to 20 heads, A	1000
(Kugler)	Various	Screw and tear-off	1, S/A or A	83
Goddard (Zalkin)	Various	Roll-on, screw or crimped	1 to 12 heads, A	500
Gravfil	—	Screw and snap-on	1 to 15 heads, S/A and A	300
	—	Roll-on	1 to 6 heads, S/A and A	120

* Abbreviations: S/A Semi-automatic; A Automatic; H Hand capper; S/L Straight line (as opposed to rotary head.)

Hart (Cozzoli)	—	Corks or stoppers	S/A and A	450
	—	Screw	S/A	—
Holland	Resina	Screw, plug, press-in or press-on	Various	450
King	CAPS 60R0	Roll-on	1 head, A	65
	CAPS 60	Screw	1 head, A	60
	CAPS 120	Screw	4 or 6 heads, A	120
	CAP 60 and CAP 120 }	Press-in or press-on	1 head or 2 heads, A	60 or 120
	RPC	Press-on	Multi-head rotary	200
	CR150	Screw or roll-on	6 heads	150
Massel (Bertolaso)	Various	Corks and stoppers	1 to 12 heads, S/A and A	366
	Various	Crown	1 head or 4 heads	100
	Omega	Cork and crown	4 heads	41
Meadowcroft	—	Crown	1, 6 or 10 heads	400
Metal Box/ MetaMatic	Various	Roll-on	1 to 12 heads, S/A and A	600
Metal Closures	Various	Roll-on	1 head, H and S/A	30
		Roll-on	1 to 18 heads, A	1000
Newman	2CT	Screw	Overbelt, S/A	100
	AC	Screw	Overbelt, S/A	100
	Presmatic	Press-in or press-on	Overbelt, S/A	100
	Auto-Presmatic	Press-in or press-on	Overbelt, S/A	100
Pacunion (Bausch & Ströbel)	KS2000	Screw	A	120
	UFVK2000	Roll-on	A	120

P

Normal and pressure seals (continued)

Supplier	Machine name	Type of closure	Details of machine*	Max. no. of seals per min.
Rockwell Pneumatic Scale (Pneumatic Scale)	Pneumacap	Screw	2 to 12 heads, A	360
Skerman	Various	Screw	—	—
Stork–Amsterdam	—	Roll-on	—	460
U.G. Closures & Plastics	Various	Roll-on	1 to 12 heads, S/A and A	600

Vacuum seals

Supplier	Type of closure	Details of machine*	Max. no. of seals per min.
Bosch (Strunck)	Rubber bungs	12 and 18 heads, A	180
Fairest Trading	Prise-off, twist-off	6 heads, A	250
Hunter	Crimped, screw lug, tumbler and press-on	1, 6 and 10 heads, H; S/A, A, and S/L	500
Metal Box	Prise-off, twist-off and P.T.	S/L 'vapour-vacuum' H; S/A and A	250, 300, 600 or 1200
Metal Closures	Roll-on	1 to 18 heads, H; S/A and A	1000

* Abbreviations: S/A Semi-automatic; A Automatic; S/L Straight-line (as opposed to rotary head); H Hand capper.

10. Processing

'Processing' includes all treatments given to filled containers; most important are the heat treatments given to many foods and drinks. The main purpose of heating is to destroy some or all of the micro-organisms; in addition, food may require cooking to make it palatable. If the time or temperature required for cooking is significantly greater than that required for sterilisation, it is usually simpler to carry out some pre-cooking in bulk before the food is bottled. In nearly all cases, therefore, the treatment in the bottle depends entirely on the bacteriological requirements. Sterilisation of filled containers by radiation instead of heat will also be considered briefly.

Some products are heat-treated before being filled and receive no subsequent in-bottle treatment. Pasteurised milk is the foremost example of this; the milk is rapidly heated to about 60°C, and rapidly cooled, thus destroying many of the micro-organisms without affecting the fresh flavour of the milk. It is then filled cold into bottles which have been sterilised by caustic washing. Other products whose flavours are rather less heat-sensitive, such as fruit juices and some wines, may be pre-heated and filled hot. Here the product is actually heat-treating the bottle and helping to ensure that it is sufficiently sterile. Finally, in special cases, aseptic packaging may be considered; this consists essentially of getting the product, the container and the closure separately to a very high degree of sterility, and then filling either hot or cold. In this way a very long shelf-life may be obtainable.

The various terms used to describe heat treatments do not have exact or universally accepted definitions, but we should clarify the meanings intended in this book:

Pasteurisation: A heat treatment, usually to less than 100°C, intended to reduce the numbers of *some* types of micro-organisms to a determined level.

Sterilisation: A treatment which aims to reduce the numbers of *all* types of micro-organism to a very low level. If done by heat, the temperature will probably exceed 100°C.

Retorting: Any heating process carried out in a pressure vessel.

Aseptic treatment: A treatment after which no growth of micro-organisms or product deterioration will occur during the planned life of the product. If the required life is very long, the organisms must be reduced effectively to zero.

It must be stressed that food sterilisation is an exact science, in which there is no excuse for guesswork. The present chapter does not pretend to be a complete practical guide to food processing, and gives only the outlines and basic principles, concentrating on those parts of the subject which most directly affect the container and closure. Some times and temperatures are quoted by way of example, but exact processing schedules must be worked out by the packer for each product and each container, if necessary in collaboration with one of the well-known food research organisations.

Finally, this chapter deals with some of the methods possible for inspecting filled and sealed containers for foreign bodies in the contents.

MICRO-ORGANISMS AND THEIR HABITS

There are many micro-organisms whose presence in food can seriously affect its safety, its quality or its keeping abilities. There are the pathogenic types, i.e. those which can multiply and cause illness if they get into the human body, and there are others which, although themselves harmless, produce organic poisons known as toxins. The most infamous example is *Clostridium botulinum*, which occurs widely in soils and on vegetables throughout the world. If sealed in an air-tight container containing a non-acid food, it will multiply and produce a deadly toxin, which was responsible for many deaths in the early primitive days of food processing.

There are also bacteria which can spoil food flavours by making chemicals such as lactic or acetic acids; and yeasts, moulds and fungi which can alter flavour and appearance in various ways.

In order to destroy a particular organism by heating, a specific minimum temperature must be exceeded, and the time required to destroy it will be reduced as the temperature is increased further. The minimum temperatures required for different kinds of organism vary widely, and there are some which can survive and even multiply at temperatures up to 80°C. Some types of bacteria produce spores, which are dormant cells with a protective coating. This enables them to withstand higher temperatures than their 'parents', and they can remain dormant until external conditions become favourable for growth, when they will germinate and multiply. Temperatures over 100°C are needed for effective spore destruction.

Micro-organisms are just like any other living beings, in that the numbers of any species do not all behave alike; in particular some will have a much greater resistance to heat than their fellows. This means that when a constant lethal temperature is applied to a colony of bacteria or spores, the weakest members (presumably the oldest ones) die quickly, and as time progresses more and more will die. But it is usually difficult to destroy the last few survivors. In other words it is difficult to achieve complete sterility, without using very high temperatures.

A graph of heating time against number of organisms surviving nearly always takes the form shown in Fig. 10.1.

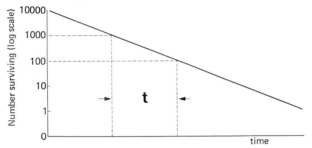

Figure 10.1 Destruction of micro-organisms at constant temperature

The logarithmic form of the relation means that in every period of time represented in the diagram by t, 90% of the organisms are destroyed. It also means that the number surviving at the end of the treatment is proportional to the number present at the beginning.

The effect of altering the temperature also reveals a logarithmic relation, as shown in Fig. 10.2a, which is taken from the classical work of Esty &

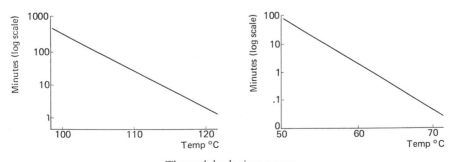

Thermal death-time curves

Figure 10.2(a). Clostridium botulinum   Figure 10.2(b). Micro-organisms in beer

Meyer* on *Clostridium botulinum*. Every point on the line represents a time-temperature treatment of equal severity, capable of reducing the number of spores to about 10^{-12} of its original value. For example, the following treatments are equivalent:

> 3·3 minutes at 120°C
> 10 minutes at 115°C
> 33 minutes at 110°C
> 100 minutes at 105°C
> 330 minutes at 100°C

It is thus clear that temperature is much more important than time in this connection, in other words small changes in temperature produce the same effect as large changes in processing time.

Food processing is made simpler by the fact that many organisms will not breed if the surrounding concentration of sugar, acid, salt or alcohol is too high. *Cl. botulinum* can only multiply or produce toxin if the pH is higher than about 4·6, so it is harmless in most acid solutions.

PLANNING HEAT TREATMENTS

Although the programmes used for some commercial heat treatments of food have been established quite empirically, it is always safer to have any new heat-treatment planned on a proper scientific basis, with the assistance of a micro-biological laboratory. The essential steps are:

1. Determine which organisms would be harmful to the consumer and to the product, and to what levels they must be reduced for safety.

2. Determine what combination of time and temperature will best produce the required result.

3. Decide what extra time must be allowed for *all* the contents of the pack to receive sufficient treatment. In effect this means measuring the heat transmission of the container and contents, and it is important that measurements are made in exactly the same conditions as will be used in practice.

4. Decide what safety margin is needed to cover any possible variations in product or in processing conditions.

5. Finally, decide what heating and cooling rates can be used.

These steps may be illustrated by considering what is required in the heat treatment of bottled beer, one of the first commercial processes to be based on the work of Louis Pasteur.

* *Journal of Infectious Diseases*, **31**, 1922.

There is virtually no danger of the occurrence in beer of pathogenic organisms, and the purpose of the heat treatment is to destroy the bacteria and yeasts which would affect the flavour, appearance and stability. Microbiologists have measured the thermal resistance of these organisms, and have determined the time-temperature combinations needed to reduce them to a sufficiently low concentration. Their results are given in Fig. 10.2b;* they show that any of the following treatments would be satisfactory:

60 minutes at 53°C
6 minutes at 60°C
0·6 minutes at 67°C
0·06 minutes at 74°C

The final choice of a particular time and temperature is largely one of practical convenience, and for beer a treatment time of about 6 minutes at 60°C is generally preferred.

Beer bottles are usually pasteurised in an upright position, by heating them with hot water sprays, and it has been found that under these conditions the temperature in the middle of the bottle lags about 10 minutes behind that at the glass surface. This simply means that in practice the bottle is held at 60°C for 16 minutes instead of the theoretical 6 minutes. (The expressions 'holding temperature' and 'holding time' are commonly used in this connection.)

Commercial equipment is available for carrying out this heat treatment under closely controlled conditions, and it is not usually found necessary to add any safety factor to the above schedule. In fact, some beer bottlers are able to manage with a shorter time or lower temperature, either because their beer initially contains fewer organisms than usual or because the beer does not need a very long shelf-life. Some brewers find it possible by very careful attention to the conditions under which the beer is made, including a special stage of filtration, to reduce the initial bacteria population to the level where pasteurisation can be dispensed with completely.

The temperature can be raised to and lowered from the holding temperature as quickly as the equipment and the bottles will allow. For small returnable beer bottles, heating and cooling rates of the order of 3 to 5°C per minute are reasonable, but it is difficult to generalise, and the matter should be discussed with the suppliers of the equipment and of the bottles.

The principles illustrated by the above example apply in exactly the same way to any food, whether solid, semi-solid or liquid. For low-acid

* After Del Vecchio, *Proc. Amer. Soc. Brewing Chemists*, 1951.

foods, i.e. meat, fish and most vegetable products, the thermal death-time curve to be used will usually be that of *Cl. botulinum*, and the time-lag in the centre of the jar must be measured carefully. For these foods, the holding temperatures required are generally in the region of 115 to 120°C, and the holding times anything from about 20 to 120 minutes, depending on the thermal conductivity of the product and the size of jar. Heating and cooling rates of the order of 5°C per minute are reasonable.

Acid foods, such as fruits, or vegetables which have been artificially acidified (e.g. pickled onions), can safely be processed at 80 to 85°C, a temperature which can be easily obtained without the use of high-pressure steam. The holding times for fruits, etc., are generally shorter than those for the low-acid foods, because the former are usually packed in such a way that liquid can circulate freely inside the jar, and so improve the heat transfer.

The simple method of calculating the holding time, by adding the time-lag at the centre to the 'theoretical' time, is slightly over-cautious. It gives the treatment necessary to deal with organisms at the centre of the container, but with liquid contents the convection currents prevent any organisms from staying there throughout the process. Also with solid foods the bactericidal effect, averaged over the whole contents, will be significantly greater than that calculated for the centre. Modern methods of computation are able to take these points into account.

PRACTICAL METHODS OF HEAT TREATMENT

The practical methods of heat-treating foods packed in glass have evolved differently for different products, mainly because of the range of temperatures required, and the variety of containers and closures used. Four methods are widely used.

1. *Overhead water spray*. This method is particularly suitable for heating narrow-neck bottles with a streamlined shoulder, because the falling water will flow quickly and uniformly down the sides of such bottles, and the heat transfer and temperature control will be efficient. Holding temperatures up to nearly 100°C are possible, but the main use of the method is for beer, which as explained above is usually heated to about 60°C. The process is always made continuous, by providing a tunnel through which the bottles pass continuously, being heated and cooled* by the same method. This method is also used for fruit products, and sometimes for wine.

---

* The same method is sometimes used just for cooling, for example, when jars of jam need to be cooled before labelling and packing.

2. *Hot-water immersion.* This was the original method of pasteurisation, and can be carried out very easily as a batch process in open pans or kettles at temperatures up to 100°C. With wide-mouth jars, a method of sealing must be used which will not allow the closures to be forced off by a pressure rise in the jar, nor any leakage of water into the jar during cooling. The water must be stirred, either mechanically or by air or steam, to obtain uniform temperature. This method is widely used for fruit, and some vegetables.

3. *Heating by live steam.* Steam is an efficient and reliable heating medium, and is widely used both in batch and continuous processes. Its rapid heating power can sometimes cause difficulty with jars sealed with vacuum-closures, as the rapid temperature rise may produce too much pressure inside the jar relative to the pressure outside the jar. Steam heating therefore is mainly used for bottles which have a pressure seal such as the crown cork; the most important application is for sterilising milk.

4. *Immersion in super-heated water.* This is the method generally used for heating vacuum-sealed jars. The jars are totally immersed in water, which is heated by steam injection. To obtain temperatures over 100°C the operation has to be performed in a pressure vessel; the pressure outside the jars can be maintained at any level required, by pumping compressed air into the space above the water.

Other heating methods. It is also possible to heat bottles or jars with hot air or by radiant heaters (usually electrical), and both methods have been used in some special applications. For accurate temperature control in the range under discussion they cannot compete with the use of water or steam. They are more useful for drying jars, or for heating them before filling.

Effect of heat treatment on bottle and closure

It is instructive to examine the main commercial processes to see what happens to the bottle and the closure.

Pasteurisation of beer. The only essential difference in the designs of various commercial pasteurisers is in the method used to carry the bottles through the heating tunnel. Some, such as the Webster Autospray, consist simply of a broad moving belt which carries the bottles, up to about 50 abreast, from one end to the other. Barry-Wehmiller use the ingenious 'walking beam' principle, in which the bottles move forward but the supporting grid does not. The grid consists of thin rails running from end to end of the pasteuriser. All the 'odd' rails, i.e. the 1st, 3rd, 5th, etc., are linked together to form one grid, and all the 'even' ones form another grid. Each grid performs a cycle of movement – up, forward, down, back – out of step with the other grid, so that the bottles are first lifted forward a short distance by the 'odd' grid, then lowered on to the 'even' grid which in its turn rises and carries the bottles forward again. It is a type of motion which is very difficult to describe in words, or to illustrate, but it works! Consequently each part of the grid stays permanently at a constant temperature, which helps thermal efficiency. The Gasquet pasteuriser built by Clayton uses a conveyor consisting of separate bottle carriers, each holding about 100 bottles. The carriers proceed to the far end of the pasteuriser, the hot end, and are then lowered vertically to another track and returned to the end where they started.

These differing methods of bottle movement are mainly concerned with obtaining the highest thermal efficiency, and they do not greatly affect the bottle.

The bottle and contents are cold, about 5–10°C, when they enter the pasteuriser, but there is no significant risk of thermal breakage when they meet the first hot water (about 40°C) because any tension produced will be on the strong inside surface the glass. As the beer temperature rises, the pressure in the bottle will rise. If the bottle is moved jerkily or unevenly the pressure is likely to become higher, i.e. nearer to the equilibrium figures given in Appendix F. The greatest stress comes at the end of the holding time, when the first cooling stage is reached. An initial temperature drop of about 15°C in the cooling water will not add excessive tension to that already being caused by the initial pressure.

In practice bottle breakage during pasteurisation is very rare, and is almost entirely confined to bottles which have been damaged in previous service or transport. If there are a few bottles to be eliminated in this way, it is obviously better that they should be broken in the washer when empty, rather than in the pasteuriser when filled and capped, and this should be borne in mind when adjusting temperatures at the two stages. The only other likely cause of breakage is insufficient vacuity due to overfilling.

Leakage of the crown corks, or other types of pressure seal, is practically unknown in pasteurising, because their leakage pressures are unlikely to fall below 7 bars.

Sterilisation of milk.   The crown cork is now used almost universally for the sterilisation of milk, so again leakage problems do not arise. In considering the pressure effects on the bottle, we must distinguish between the batch process and the continuous process.

In the former, the bottles are loaded into a large horizontal retort through an end door, and heated by steam injection. (It is essential to allow the steam sufficient time to displace all the air from the retort before it is finally sealed, but this is to obtain the correct final temperature, and does not affect the bottles directly.) The usual sterilising temperature is 115°C, and the holding time similar to that for beer bottles. The internal pressure in the bottle depends on the vacuity and the temperature rise; for a bottle allowing 10% vacuity, and a temperature rise of 45°C above the filling temperature, the internal pressure may be of the order of 2 bars, or higher if the bottle is overfilled. So long as there is high-pressure steam outside the bottle, the net effect on the bottle is negligible, but when the steam pressure is released at the end of the treatment, the full effect of the bottle pressure is realised suddenly. Cooling is usually started by spraying the bottles with the hot water which has condensed in the bottom of the retort, gradually adding cold water to it to lower its temperature steadily. It is important to see that the simultaneous release of steam pressure and the first cooling shock of the water do not combine to cause excessive tension in the glass.

In a continuous pressure steriliser it is necessary to provide a pressure seal through which the bottles pass, and in the Webster 'Pioneer' this is achieved in the manner shown in Fig. 10.3.

The principle of this machine is appealingly simple: the seal consists of a vertical column of water which is balanced exactly by the pressure of steam in the sterilising chamber. This method lends itself to very precise temperature control, and the only possible disadvantage is that it is not so easy to alter the sterilising temperature at will. It is possible to build a steriliser on this principle with sufficient 'spare' height to permit alteration of the temperature, but this has not been found necessary with milk.

There is an important difference between the two methods of milk sterilisation, in regard to the pressure in the bottle. In the continuous steriliser, when the bottle leaves the steam chamber, the external pressure (now provided by the water) falls slowly and steadily to atmospheric as the bottle rises in the water seal, being cooled as it goes. This means that the *excess* pressure inside the bottle will probably not rise to more than about 0·5 bars.

In the form of steriliser illustrated in Fig. 10.3 the bottle receives little agitation as it is processed. It travels in a pocket (the main belt may have as many as 20 bottle pockets across its width) which is turned through 180° at top and bottom of each pass. If more agitation is needed, a sinuous bottle path can be provided which gives continuous gentle agitation of the bottles. This helps to produce a stable, uniform product and prevents deposition of sediment on the inner container surface. Machines have

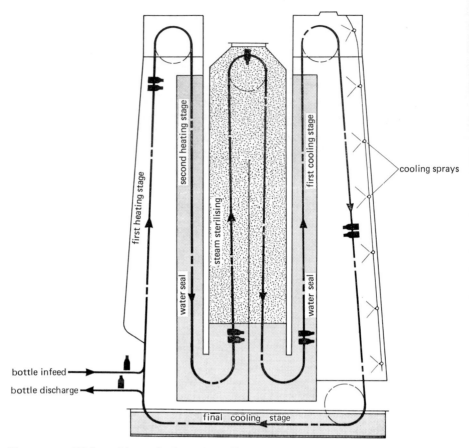

Figure 10.3. Webster Pioneer hydrostatic steriliser

also been made which rotate the bottles continuously while being processed.

No reference has been made so far to the outputs possible for any of the continuous pasteurisers or sterilisers. All that need be said here is that there is no practical limitation, as plant is available which can cope with the output of the fastest filling machine.

RETORT PROCESSING OF VACUUM-SEALED JARS

The effect of pressure on the jars can be entirely neglected when using vacuum closures; the important question is the effect of pressure on the closures themselves. As explained in Chapter 9, most vacuum closures are not designed to retain their effect when the internal pressure exceeds the external. Prise-off closures will be forced off if the internal pressure rises too high, and vacuum screw closures will leak. In all these cases the closure is controlled throughout the heating and cooling by applying extra air pressure to the top of the retort, adjusted to a value which will always keep the pressure outside the jar greater than that inside it.

Crimped-on closures such as the Omnia behave slightly differently. When the internal pressure rises, the cap lifts slightly, and it can be relied on to re-seal properly when the excess pressure has been released, provided that there is no solid obstruction between the sealing gasket and the glass. This action can be employed, with discretion, to increase the degree of vacuum in the final pack. The method is to delay applying the extra air pressure, until the caps have lifted and released some of the air in the jar. The compressed air is then applied, and the closures return to the sealed position.

The retorts used for processing glass jars are steel cylindrical pressure vessels, usually vertical (i.e. with the axis of the cylinder vertical) but larger models may be of the horizontal type. A battery of vertical retorts supplied by Mather & Platt is shown in Fig. 10.4.

The essential services and fittings required for water-filled retorts are:

1. Water supply and drain.

2. A means of injecting a controlled supply of steam at the bottom of the tank.

3. A method of stirring the water. In vertical retorts this can usually be done by blowing compressed air in at the same place as the steam. In horizontal retorts, a circulating pump for the water is likely to be needed. This part of the system and the steam supply need to be designed carefully, to ensure uniformity of temperature at all stages, without moving the water so violently that it will disturb the jars.

4. A controllable supply of compressed air to the top of the tank, to maintain the top pressure at the required value.

5. A means for cooling the water in the tank at the end of the process. If a circulating pump is used for mixing, it is easy to 'bleed' cold water into the system progressively. In vertical retorts it is usual to provide a set of water inlets round the top of the retort.

6. Instruments for indicating and controlling temperatures, pressures and water level.

It saves processing time if the starting temperature of the water is about the same as that of the jars when they are inserted. (If the water is much hotter than the jars, the closures may leak before the processing starts.) Normally the jars are loaded in open baskets, with sufficient openings or perforations to allow good water circulation.

Figure 10.4.   Mather & Platt vertical retorts

The pressure at which the retort has to be operated depends on how much the pressure will rise inside the jars, and this depends on vacuity, expansion coefficient of the product and the pressure of air or gas originally sealed in the jar. A few tests will soon provide the answer, but for most products and conditions a retort pressure of about 2 bars might be suitable for a process temperature of 115°C.

Cooling rates should be determined by experiment, but, as mentioned earlier, 5°C per minute is a reasonable rate. If possible, processed jars with tinplate caps should be removed from the autoclave at a temperature of about 50°C, so that the remaining heat will be sufficient to dry the caps and prevent rusting.

Hot filling

As referred to earlier, a good method of pasteurising fruit juices and similar products is to heat them rapidly by passing them through a heat exchanger immediately before filling. This method relies on the heat of the juice to help destroy any organisms on the inside of the pack. Its main advantage is that it avoids the development of off-flavours and discoloration which may result from prolonged heating of the juice in a bottle.

For efficient pasteurisation – and to avoid the risk of thermal shock to the container – the bottles need to be pre-heated to within about 60°C or less of the filling temperature. Most fruit juice bottles are sealed after filling by the vapour-vacuum method (see Chapter 9.) This minimises the risk of oxidation, which could cause loss of vitamin C. Finally, the bottles are cooled as quickly as possible by an overhead water spray or fine mist nozzles.

Aseptic packaging

Aseptic packaging is in use in the United Kingdom for filling 'long-life' milk into paper cartons and in the USA for filling fruit juices into bottles. Apart from the sterile filling of drugs into ampoules and small bottles, there has not been a great deal of commercial interest in aseptic bottling in Britain, no doubt because the process tends to be more costly and difficult to control than conventional pasteurising or sterilising methods. It involves heating the product to a very high temperature, possibly around 150°C for a short time, then cooling it and filling it into organism-free containers. One UK system has been developed in collaboration with the National Institute for Research in Dairying for the aseptic filling of milk and other dairy products in bottles, and although there are no signs of the process supplanting pasteurised milk in the foreseeable future, this could be the forerunner of similar processes for other foods and drinks.

Sterilisation by radiation

A modern method of destroying micro-organisms at room temperature (not to mention macro-organisms) is to give them a heavy dose of ionising radiation, either with gamma radiation from a radioisotope or a beam of ionising particles from an electrical machine such as a linear accelerator.

This method still presents many unsolved problems in the treatment of foodstuffs, but it may be useful for such things as heat-sensitive pharmaceutical products.

Glass containers can be subjected to this treatment without any risk of damage. If ordinary colourless glass is used, the radiation turns the glass a brownish colour, not unlike that of amber glass, but does not affect its strength or other properties. The colour change could in some circumstances be an asset, since it would provide irrefutable evidence that the container had been sterilised. If necessary the irradiated glass can be restored to its original colourless condition simply by heating to between 150 and 200°C.* When amber glass is irradiated, there is a much less noticeable colour change.

The radiations used have good powers of penetration, and there is no difficulty in penetrating glass container walls. Soda-lime glass has just about the same penetrability by radiation as the equivalent thickness of pure aluminium metal, and it is much more transparent to radiation than other metals such as steel.

Inspection of filled containers

There are a number of different inspection functions which a bottler may apply before the filled bottles pass to the labelling operation, although some of them can be done just as effectively after labelling and can then be combined with a check that the label itself has been correctly applied.

---

* If irradiated glass is heated in the dark, it shows the interesting phenomenon of thermo-luminescence, i.e. it gives off a blue light while the brown colour is disappearing. This is the way in which the glass disposes of the energy it has absorbed from the radiation.

INSPECTION FOR CONTAMINATION

The first inspection that may be required is for any foreign matter in the contents of the bottles. Obviously, a thorough visual inspection is possible only for clear transparent products. The normal 'manual' method of doing this is to invert the bottle and watch for any foreign bodies rising or falling. British Miller Hydro supply an in-line machine, called the Autospector, which performs the mechanical part of this operation automatically. Each bottle is inverted as it moves in front of an illuminated screen, in view of a 'sighter'. Suspect bottles can be removed at the sighting position, or the operator can place a pre-gummed reject tag on the bottle. Machines with a single sighting station can handle up to 60 bottles per minute, but there are also two-sighter units capable of dealing with 120 or 240 bottles per minute.

Several companies have tried with varying degrees of success to automate inspection for product contamination, and to deal also with opaque products as well as transparent ones. It is obviously very difficult to design a commercial machine which will work with a wide range of containers and different types of contamination, but some packers will remember with embarrassment or amusement an apparently magical 'black-box' which was convincingly demonstrated at exhibitions some years ago, but which did not correspond to any known scientific principle. There seems to be more scope for developing machines for dealing with a single type of container, product and contaminant. In very recent years, however, electronic techniques generally and medical X-ray methods particularly have developed very rapidly, and we may be on the brink of a new inspection era. The only other question to consider then is whether it would not be better to spend the money and effort on preventing the contamination at source?

For ampoules and vials, the process of inspection has been successfully mechanised in the Emhart Autoskan machine. This is designed for the detection and rejection of parenteral solutions containing particulate matter of about 40 microns or more in size (although in some circumstances even smaller particles can be detected).

The ampoules or vials are individually fed into an 8-station turret, where they are rotated at various speeds at each of three different stations. The purpose of rotation is firstly to set the fluid and any particulate matter in motion, and secondly to 'clean' the interior wall of foreign particles. At the fifth station the container is examined by a TV camera as a moving liquid in a stationary container. A master 'picture' is taken and recorded in a memory unit. Then four secondary 'pictures' are taken and each is compared with the master. If any one of the four is not identical

Q

with the master, an error signal is generated. (The picture of a contaminated container would not be identical to the master since the moving particles would be in different positions in any one or more of the secondary pictures.) Stationary particles such as dust, or printing, etc., on the outside of the container are disregarded by the system. Rejection can be specified for either one, too, three or four errors per test, and the entire test procedure can be repeated up to four times. The sixth and seventh turret positions are the accept and reject stations. If, for some reason, a container should pass both these stations, the machine will stop automatically when it reaches station eight. Throughput is up to 60 containers per minute.

FILL LEVEL INSPECTORS

A means of checking bottles for the correct fill height is now becoming more and more important, not only to satisfy legal requirements as to the quantity of product in the pack, but also to prevent over-filling, which can be quite expensive in terms of 'lost' product.

Machines for checking the fill level automatically usually operate by means of a beam of radiation. In the Hytafill 1000 level inspector, made by General Electric (USA), a fine beam is projected through the containers at a given level and the unit can be set to detect either under-fill or over-fill. When used for the inspection of bottles the Hytafill 1000 is usually also provided with a cap detector, and can operate at up to 650 bottles per minute.

Another range of fill-level detectors, called Filtec, is made by Industrial Dynamics Company, of California, and supplied in the UK by Metal Box. In addition to the inspection of fill level, other optional inspection functions can be provided for missing or tilted closures, missing or improperly applied labels, and such specialised functions as low foam in bottled beer.

Fill level inspectors

Supplier	Name of machine	Principle	Max. input speed per min.
Glenhove (Datz)	—	Electronic	666
Landré-Mijnssen (General Electric USA)	Hytafill	X-Ray	650
Metal Box/MetaMatic	Filtec	Gamma Ray	800

11. Labelling

The term labelling is usually taken to include all forms of information or decoration placed on a pack after it has been filled and sealed. With glass containers there are in addition two ways of labelling before filling; by embossing the glass surface, or by printing directly on the glass surface. In this chapter we will consider all these methods.

Labels may have to fulfil three functions:

to identify the product and the packer,
to supply any information needed by the user or required by law,
to decorate the pack.

Identification of the product is straightforward, and for glass packs it may even be unnecessary, as the user can see for himself what the bottle contains. For example, there is no need for a dairy to state on its bottles that they contain milk. But there are, of course, ever-increasing legal requirements about product identification and statement of constituents. Identification of the packer is also helped by the use of a well-designed glass container: no one has to look for a label on the Coca-Cola bottle to find out what the liquid is and who made it. So if the filled bottle is, in effect, its own identifying label, and if no other information needs to be given, the situation may be met most simply and most cheaply by embossing the packer's name on the bottle.

Decoration of the pack is not done just to make it look pretty; this is done primarily to assist in selling it, and labels should be designed with this in mind. It is also important that the design of the bottle, closure and labels should be considered as a whole. If a label printed direct on glass is to be used, its colours should be chosen in relation to the colour of the glass, or rather the apparent colour of the glass when it is filled with the product.

Labels will be considered in three groups: first those printed on glass, secondly plain paper labels, which are applied with a wet adhesive, and finally all those which can be stuck to glass without separate adhesive.

Labels printed on glass

VITREOUS ENAMEL

Although a relatively costly method of labelling, vitreous enamelling is often of particular value for multi-trip bottles, as it can be relied upon to last the life of the bottle without any deterioration. Each colour is applied to the bottle through a 'silk-screen' stencil, a process used for many types of printing, in which the flow of enamel through the stencil openings is controlled by a fine mesh of nylon or metal. Modern enamels set hard almost as soon as the stencil is removed, and several different colours can be applied in rapid succession. The operation is completed by heating the bottle to between 500 and 600°C, at which temperature the enamel is fused to become an integral part of the bottle surface. The heating and cooling have to be carried out under closely controlled conditions, and it is necessarily a job for the bottle maker, not the packer.

The range of colours is as wide as with any other printing process, and designs incorporating fine-line and tone-work can be printed in one to four colours.

Vitreous enamel can also be applied by a transfer, which usually consists of a collodion film on a paper backing; the collodion is burnt off during the firing of the enamel. The process is more expensive than silk-screening, but it may be useful for labelling parts of bottles which are not accessible to a silk screen, or if more than four colours are required.

COLD COLOUR PRINTING

Considerable progress has been made in the development of synthetic colours which do not require firing after printing on glass. They are based on epoxy resins with various catalysts to harden them, and they can be applied by silk-screening or similar methods. Their appearance is equal to that of the vitreous enamels, and they withstand a fair amount of abrasion without damage.

Methods also exist for printing synthetic colours on cylindrical glass surfaces by a simple rotary printing process. The density of colour obtainable is not very great, but its undoubted cheapness has made it acceptable in some situations where appearance is not of primary importance.

Paper labels

SELECTION OF PAPER FOR PLAIN LABELS

Sticking a paper label on a bottle by hand is a very easy operation, and almost any paper would be satisfactory. For a label whose appearance is commercially important, and which is to be applied mechanically, four factors determine the choice of paper:

the colour and appearance of the surface, and its suitability for the chosen printing method,

the physical behaviour of the paper during and after application,

the chemical properties of the paper (occasionally),
the cost (always).

The first of these factors need not concern the packer unduly, as it is primarily the printer's job to select a paper which will be dimensionally stable on his printing presses, and suitable for the inks and varnishes he may employ. Uncoated papers, i.e. those with a smooth surface obtained solely by rolling ('calendering'), are usually too absorbent for label printing, and generally some type of surface coating is chosen, usually based on china clay. Some types of paper are liable to discolour when exposed to sunlight, especially those made from wood pulp by a mechanical method. If discoloration must be avoided, paper made by a chemical method will be needed, and as this contains no 'mechanical' wood pulp, it is sometimes described as 'pure' or 'wood-free'.

The behaviour of the paper during label application is very important, and becomes more so as one goes from manual to semi-automatic operation and thence to high-speed automatic methods. The back of the label needs to be sufficiently rough and absorbent to provide a good key for the adhesive, and the paper should not be so thin that the adhesive will penetrate to the front, or so translucent that the printing will be visible from the back. Occasionally paper labels are printed on both sides, which sets the paper maker a nice problem in compromise. For most labels, a weight (called 'substance' in the trade) of between 70 and 90 g/m²* is usually found suitable.

The vital property of the paper during its application to the bottle is its stiffness: it must be stiff enough to be handled by the machine without crumpling, and flexible enough to be bent to the shape required to fit the bottle. It is not possible to measure stiffness directly, so it cannot be included in a label specification, but the three properties which determine

*Grams per square metre is often incorrectly abbreviated as 'gsm' in the paper and board industry.

stiffness – weight, caliper (i.e. thickness) and type of surface coating – can and should be specified.

If the label paper is given a heavy varnish coating, or lamination of aluminium foil, its stiffness will be considerably increased, and the paper weight may have to be decreased to partially compensate. Foil labels usually consist of a foil about 9 microns thick glued to a paper of 40 to 60 g/m²; even so such labels are considerably stiffer than ordinary paper labels of 70 to 90 g/m², and need special treatment and stronger adhesives. Sometimes it may be possible to overcome these difficulties by the use of vacuum-metallised papers instead of foil/paper laminates. These have the appearance and lustre of foil, but are usually cheaper and easier to apply.

In all papers, most of the fibres lie in one direction, 'the grain direction', and because of this the paper tends to curl when wetted on one side, as explained later. Labelling would be easier if curling did not occur, but no paper is immune from it. The amount of curl varies, depending on factors such as the average fibre length, and when choosing a label paper it is advisable to check that its curl will not be excessive for the intended method of labelling.

The chemical properties of the paper may need considering if a strongly acid or alkaline liquid is liable to be spilt on the label at any time. Ordinary paper has a pH value of 5·5 to 7, i.e. it is chemically nearly neutral, but if necessary paper can be provided with a higher or lower pH, to minimise any discoloration or loss of adhesion. When this is done, the printing inks, varnishes and adhesives may have to be chosen accordingly.

Perhaps the most important property of a labelling paper is uniformity. Paper is, of course, made from natural raw materials, and it is difficult to avoid some variations, but unless they are kept under close control, variable results on high-speed labellers are likely. Apart from natural variations, the label maker should not be allowed to change the type of paper or method of manufacture without consultation. A clear label specification is greatly in the interests of efficient operation; it should cover all the materials used, the methods of printing and cutting, and the grain direction required.

PRE-GLUED PAPERS

The use of pre-gummed paper for luggage labels and postage stamps is well-known, and exactly the same material can be used for labelling bottles. The paper is coated with gum arabic or dextrin glue and allowed to dry; before applying the label it is moistened with water. Pre-gummed labels provide a simple and clean method for the hand-labelling of small batches of bottles, but they cannot be applied by machinery.

A more useful material, which can be applied by machine, is the heat-sensitive paper which is pre-coated with thermoplastic resin. When the paper is heated the resin becomes tacky and forms an effective bond with glass. One type of paper suitable for glass is known as 'Delatac' and is supplied by Samuel Jones, who will advise on its handling and printing. There is no general limitation on printing methods for 'Delatac' but some types of ink are liable to soften the resin, and if a varnish is used by the printer it must have a high enough softening temperature not to soften when the adhesive is activated. Heat-sensitive labels have very good adhesion to glass, and they are much more difficult to wash off than labels which have been stuck on with water-soluble glue. If, therefore, they are considered for re-usable bottles, the packer should make sure that his washer will be able to deal with them.

Thermoplastic-coated labels are usually supplied cut to size, just as is done with plain paper labels, but there are also labelling machines which take the labels in the form of a continuous reel, and automatically cut off the rectangular shape intended for each individual bottle.

'Pressure-sensitive' paper labels are also available, made from paper with a permanently sticky backing, protected by a separate removable paper backing. The material itself is generally more expensive than pre-gummed or heat-sensitive paper, but the application is so simple, requiring neither glue, water nor heating plate, that the overall convenience and costs may be attractive in some situations. Their use has been steadily increasing recently, particularly for small bottles which are difficult to handle through conventional labelling machines.

Two types of pressure-sensitive coating are available; a 'permanent' one, which means that the label cannot be removed without tearing the paper, and a 'temporary' one, which enables the label to be peeled off at any time. Pressure-sensitive labels can be supplied in sheet form, for hand labelling, or in reel form for automatic labelling. In the latter case, the backing paper is continuous but the labels are already punched out to size.

Transparent cellulose film has occasionally been used to make pressure-sensitive labels for glass. It is supplied in continuous reels, and in this form does not require a protective backing. It is not too easy to cut into accurate rectangular pieces for the individual labels, and is probably best suited for a complete wrap-round label, for which a variable amount of overlap may be tolerable.

Design of paper labels

It takes only a few moments' observation in a supermarket or chemist's shop to realise that infinite variety of label design is possible, and that there are endless permutations of shapes, positions and numbers of labels, not to mention the many different appearances provided by different printing methods, foil lamination, embossing and other techniques. No attempt therefore will be made here to tell the packer how to design his label array; so much depends on the technical and commercial needs, the individual ideas of the packer, and the permissible labelling cost. A number of technical considerations must, however, be borne in mind throughout the design process, and twelve points are listed below.

DESIGN IN RELATION TO LABEL MANUFACTURE

1. The designer should if possible know the method of printing likely to be used, so that he can take advantage of its capabilities and allow for its limitations. As a rough rule, small quantities of labels will probably be printed by ordinary letterpress; for very large quantities gravure printing would be economic; between the two, lithographic methods are appropriate.

2. Dimensional accuracy is very important for mechanically applied labels. This factor is becoming more and more vital as labelling speeds increase. A dimensional tolerance of $\pm 0 \cdot 4$ mm would be considered fairly normal, but today some fast labelling machines require an accuracy of $\pm 0 \cdot 2$ mm.

3. If a plain border or margin is left round the edge of a label it should be at least 3 mm wide, otherwise any slight inaccuracy in cutting the label may be emphasised.

4. The grain direction of the paper is important for mechanical labelling. When paper is wetted on one side (with water or water-based glue) the wet side expands much more *across* the grain, with the result that a square piece of paper will curl into part of a cylinder (wet outside), with the grain running along the axis of the cylinder. It has nearly always been found that with plain paper labels best results are obtained with the grain horizontal (i.e. at right angles to the axis of the bottle). This is a fairly general rule, applying also to complete wrap-round labels, but there are a few exceptions where an unusual bottle shape or label shape makes a vertical grain preferable. The most important point is that the grain direction should not be allowed to vary.

The effect of grain direction is different with heat-sensitive labels: the adhesive takes some time to form a firm bond, and during this time the edges of the label are liable to lift away from a curved glass surface. Fortunately, the tendency to lift is much less in a direction across the grain, so the paper can best be fitted to a curved surface by having the grain direction as shown in Fig. 11.1. If the glass surface is flat, and the label long and thin, the grain should run parallel to the longest edge.

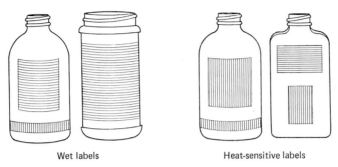

Wet labels   Heat-sensitive labels

Figure 11.1.   Recommended grain directions for paper labels

DESIGN IN RELATION TO THE BOTTLE

5. A paper label can only be fitted smoothly to a surface which curves in a single direction, or which is part of the surface of a cone (in addition, of course, to perfectly flat surfaces). Parts of the bottle which curve in two directions, e.g. the lower part of the shoulder of a 'cylindrical' bottle, should therefore not be labelled. As explained in Chapter 3, it is very difficult to make flat panels of bottles perfectly plane, and some of them may sink or bulge slightly. Large panel labels might therefore be liable to wrinkle slightly. This is not likely to be troublesome if the labelling machine can stretch the label to fit as it presses into position, but even so it is advisable to keep panel labels reasonably small.

6. For machine labelling, the labels should be placed in positions where the machine can effectively 'wipe' them. (This term is explained later in this chapter.) Concave surfaces should be avoided for this reason (and also because of point No. 11 below).

7. Labels should preferably not be placed in positions where they are liable to get rubbed or scraped in subsequent handling. Many bottle designs provide a recessed area to protect the label or a raised protecting ridge. If a label edge coincides with a corner or sudden change of direction of the glass, it is liable to be damaged or pulled away from the glass.

8. Labels should not be placed with their edges on surfaces where the radius of curvature is small, since the springiness of the paper may make the edge lift away from the glass before the adhesive has set. This applies particularly to heat-sensitive paper. Label shapes with sharp points are best avoided, especially if the labelling machine only puts glue on part of the label.

DESIGN IN RELATION TO LABELLING MACHINE

9. Some labelling machines cannot deal satisfactorily with a glued label which extends more than halfway round a bottle, because of the method of wiping used.

10. If wrap-round labelling is employed the underlapping end of the label should be kept free of print or varnish.

11. It should be remembered that a wet label shrinks as it dries, and a heat-sensitive label expands as it picks up moisture from the air after being dried by heat. Both facts are reasons against labelling a concave surface.

12. If circular labels are used, means must be provided for placing them at the correct angle in the label magazine. The usual method is to cut a notch in the edge of the label, which fits over a rib in the magazine.

Adhesives for plain paper labels

The glass packer does not need to be an expert on adhesives, apart from realising that there is a very wide range available, out of which there may be several suitable for his particular label and machine, and his required production speed.

In recent years, two factors have combined to make the choice of a suitable adhesive increasingly important. The first is the development of new, high-speed labelling machines, and the second is the general use of surface treatments for bottles, as discussed in Chapter 2. Glass manufacturers have worked closely with suppliers of adhesives and labelling machines to determine which adhesives are suitable for any combination of surface treatment and labelling process, so there should not be any difficulty in getting appropriate advice.

Adhesives for glass labelling will usually be chosen from one of the following groups:

Starch. These should strictly be called pastes, as in their simplest form they consist of starch products dispersed in water.

Dextrin. Water-based brown adhesives made from a conversion product of starch (the conversion is by acid treatment and heat). Dextrin glues were once the most popular types for glass labelling, but have recently tended to be replaced by Jelly Gums (see below).

Jelly Gums. Made from converted starches, modified to give a range of operating characteristics and of jelly-like consistency. They perform well with most types of surface treatment.

Casein. This is made from the curd of soured milk. It can provide a bond which will survive freezing and works well with most coatings, but is sometimes less easy to handle than Jelly Gum.

Synthetics. There is a wide range of adhesives made from synthetic resins. For labelling, the most common types are polyvinyl acetate (used more on plastics than on glass containers), and hot-melt (based on solid thermoplastics), which are sometimes used for all-round labelling and top-strapping.

Vegetable Gums. The natural vegetable gums (e.g. gum arabic) are mainly used for pre-gummed labels.

When considering the adhesive to be used in a particular situation, four basic points should be established:

1. The method of labelling and the speed of operation. For hand-labelling a slow-setting glue is needed; a high-speed machine will work better with a fast-setting one. Some machines pick each label out of the label magazine with adhesive-coated fingers or rollers; in this case the adhesive must therefore have a good 'initial tack'.

2. The surface treatment of the glass, if any, and the conditions prevailing at the time of labelling, e.g. whether the glass is hot, wet or greasy. If the labelling surface is stippled rather than plain, this has to be taken into account, also any peculiarities about the shape of the labelling surface.

3. The type of label. In particular if the label is unusually stiff, e.g. a laminated or heavily varnished one, a stronger glue is needed, but not so strong that it cannot be removed from a returnable bottle. With laminated labels, it is also important that the glue used to stick the label to the bottle does not upset the adhesive used to stick the foil to the paper. (Preferably the laminating adhesive should continue to hold foil and paper together when the label is washed off a returnable bottle.)

4. The conditions of use of the labelled bottle. Handling with wet hands, storage in refrigerators or steamy bathrooms, etc., indicate the need for waterproof glue. For returnable bottles it is possible to obtain a glue which is insoluble in water but soluble in caustic detergent solution.

Applying plain paper labels

A wide choice of labelling machinery is available, ranging from simple glue applicators for hand labelling, to very high-speed automatic units. Speeds of up to 1000 bottles per minute can be obtained with some single automatic units.

Hand labelling is often the most economic way of dealing with small quantities of bottles on an intermittent basis. Usually it is found best to have a small team, with one operator putting the glue on the labels, and the others applying them. Such a team, using a simple glue applicator, might label about 40 bottles a minute.

If a labelling speed of this order (40 per minute) is required all the time, instead of intermittently, a semi-automatic machine will usually be indicated. Only one operator will be needed, to place and remove the bottles. Speeds of up to about 60 per minute have been obtained in good conditions.

The use of fully automatic machines does not mean that the packer can just switch them on in the morning and forget about them until time to switch off. For efficient operation, a high standard of supervision and maintenance is needed, and also a high standard of planning to ensure that the dimensions of bottles, labels and the mechanical parts of the labeller are correctly interrelated. The author has found widely differing views among packers about the efficiency of various automatic labellers, and probably most of these differences can be attributed to the way in which the machines are used rather than to the machines themselves. An automatic labeller is a precision machine, and it should be treated as such. It should be remembered that most glues are not ideal lubricants for high-speed machinery.

BASIC OPERATIONS IN GLUE LABELLING

Four separate operations can be distinguished in any semi-automatic or automatic labeller:

Pick-up: extraction of the label from the label magazine.

Gluing: application of glue to the label, or occasionally to the bottle.

Transfer: placing the label in position on the bottle, or occasionally vice versa.

Wiping: pressing the label into its final position by pads, brushes or rollers.

Pick-up.   In all machines the labels are held in a stack in a magazine, and the machine picks them out one at a time. In one group of semi-automatic machines the pick-up is made by bringing a perforated plate close to the label stack, while a vacuum pump draws air through the perforations. Sometimes a small air jet is used at the side of the stack to help separate the end label from the others. The printed side of the label goes towards the perforated plate, so that the plain side is exposed for coating with the glue.

The other method of pick-up uses the adhesive itself. Two picker arms or a slotted picker plate, which have already been roller-coated with glue, are pressed against the unprinted side of the label, and withdrawn holding the label.

Both methods of pick-up are shown in detail in the section on semi-automatic machines (Figs. 11.2 and 11.3.)

Fully automatic labellers can also employ either a vacuum pick-up or glue pick-up system, although their modes of operation differ from those of the semi-automatic machines. A few typical examples are described in the section on automatic glue-labelling machines.

Gluing. After a vacuum pick-up, the label is passed across a glue-coated roller, thus covering the whole of the unprinted side with glue. If for any reason all-over gluing is not required, the vacuum plate can be shaped to hold parts of the label away from the roller. With the adhesive pick-up method, the label obtains its glue at the time of pick-up. With some labellers the picker arms are made to slide outwards across the back of the label, to increase the glued area.

One of the essentials of good labelling is to obtain a uniformly thin film of adhesive on the label. In all machines the glue is supplied by a roller revolving in a bath of glue; from the first roller it is usually transferred to a second roller or rotating belt, and thence to the label or the picker mechanism. The glue rollers must be in good condition and accurately cylindrical, and if a scraper blade is used to control the film thickness it must be straight and accurately set.

Transfer. If the label is picked up by vacuum, the same vacuum head usually holds the label throughout the gluing operation, and then places it in position on the bottle. In some fully automatic labellers, however, the first vacuum head passes the label on to a second one for coating with glue and application to the bottle.

With a glue pick-up method, there are several methods of transferring and placing the label. In the simplest system the two picker arms hold the label just in front of the bottle (they are at this stage between the label and the bottle), then a third arm moves forward between the two picker arms, and pins the label against the bottle while the picker arms withdraw sideways. In more complex machines there may be an intermediate transfer stage, e.g. from picker plate to transfer arm and thence to a delivery arm or drum; vacuum may be used to assist label control in these stages.

With cylindrical bottles it is sometimes necessary to place the label at a particular point in relation to an embossed decoration on the glass, or to the position of the bottle seams. There are several labelling machines which can do this. Usually the bottle is made with a small indent or protrusion (a 'spotting bar') near its base, and the bottle is rotated until the spotting bar registers with a probe on the labelling machine.

Wiping. The transfer stage usually ends with the label partly stuck to the bottle, but with its edges free; the purpose of wiping is to stick the edges down smoothly and securely. A good wiping action follows the same principle as required in wallpapering, i.e. it sticks the centre down first, and then works out to the edges. If the wiping is done with a single resilient pressure-pad, the pad should if possible be shaped so that it contacts the centre of the label first.

SEMI-AUTOMATIC GLUE-LABELLING MACHINES

Two examples of semi-automatic labeller are the Purdy 'Universal', and the 'Pony Labelrite' made by Whitehall Machinery. In both of these the bottle is laid horizontally in the machine, and the label or labels (up to three or four) are automatically stuck to the upper side of the bottle. These machines can thus only label one side of the bottle at a time, but they can apply a wide label to a round container, which is subsequently wrapped round and overlapped at the back.

The Purdy machine uses a glue pick-up method; Fig. 11.2 shows three labels in position on the picker arms, having been collected from the overhead magazines. The placing arm is about to swing down between the picker arms, and press the labels on to the bottle.

The vacuum pick-up of the Pony Labelrite is shown diagrammatically in Fig. 11.3, also the twin-roller method of feeding the glue. The thickness of the glue film is controlled by adjusting the pressure between the rollers.

AUTOMATIC GLUE-LABELLING MACHINES

Automatic machines vary considerably in their modes of operation and it is not very useful to try to classify them in definite groups. The lists at the end of this chapter will give a general idea of what is available, and here we will only mention briefly a few of the more interesting or important types.

One general subdivision could be between 'straight-line' machines, in which the bottles travel straight through on a continuously moving

Figure 11.2.   Purdy Universal semi-automatic labeller

conveyor, and those in which they are conveyed in a circular path by a rotary turret. An example of the former is the Weiss RO-300, which employs a vacuum pick-up and transfer system. The labels are drawn from a stationary label magazine by a rotating vacuum cylinder and passed to a large rotating transfer drum, which is also under vacuum. The drum takes the labels to a gluing station, where they are glued by a roller, and then carries them round to meet the bottles. The moment the container touches the label, the vacuum in the relevant section of the transfer drum switches off and compressed air aids the mechanical transfer of the label to the bottle. Throughputs of up to 600 per minute can be obtained.

A glue pick-up system is used by the Morgan Fairest 400 straight-line machine (Fig. 11.4), which can apply single labels to cylindrical bottles at speeds of up to 400 per minute. Labels are removed from a stationary

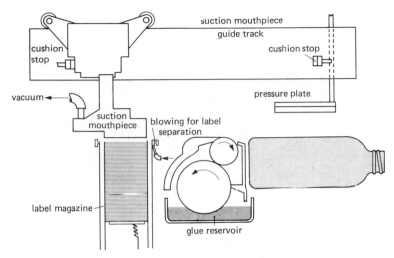

Figure 11.3.   Pony Labelrite semi-automatic labeller

magazine by the pre-glued faces of a rotating pick-out unit, then mechanically transferred to a continuously rotating transfer drum, from where they are picked up by the bottles as they roll past the drum.

Similar basic principles are employed in a number of machines which carry the bottles in a rotary turret, development of which has been particularly rapid during the past ten years. The earliest machines of this

Figure 11.4.   Morgan Fairest MF400 automatic labeller

type had oscillating label magazines working in conjunction with a glue-roller pick-up system, but most of the latest rotary labellers now employ stationary label magazines – a feature which contributes to higher operating speeds and allows the label supply to be replenished without stopping the machine.

The operating mechanism of the Jagenberg Solar range of high-speed rotary labellers is shown diagrammatically in Fig. 11.5. Label pick-up is by means of rotating glue plates or 'segments' which are mounted on an independently rotating turret next to the bottle turret. The plates first pass over a glue roller and are then carried round to the label magazine. After picking up a label, they approach a gripper cylinder which takes the labels and transfers them to the bottles.

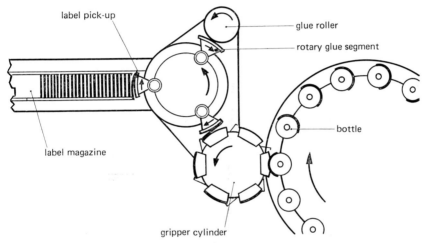

Figure 11.5. Jagenberg Solar labelling mechanism

A somewhat similar arrangement is employed in the Krones Solomatic rotary body and shoulder labeller, seen in Fig. 11.6, although in this machine the glue plates make an oscillating movement instead of a rotary one. Throughputs of more than 800 bottles per minute are obtainable on this and other rotary labelling machines.

In addition to body and shoulder labelling, many straight-line and rotary labellers can be equipped for the simultaneous application of back and neck labels, or wrap-around labels. Some will also wrap decorative foil neatly over the top and round the neck of the bottle, giving a luxury look not only to expensive wines but also to many less noble beverages.

An alternative, although slower method of applying a complex array of labels is seen in the intermittent-motion Purdy Linamatic 65 (Fig. 11.7.) As the bottle is carried through the various stations by the intermittent conveyor belt, it is steadied by the overhead assembly which moves with

R

Figure 11.6.   Krones Solomatic high-speed rotary body and shoulder labeller

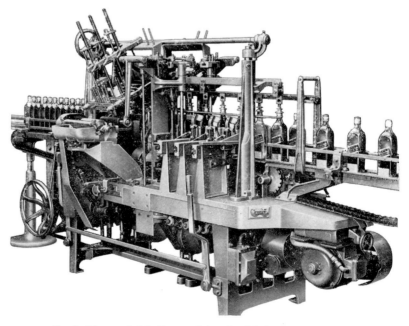

Figure 11.7.   Purdy Linamatic labeller, applying five labels in one operation

it. The label pick-up and transfer is a more advanced version of the system described earlier for the Purdy semi-automatic; the difference is that the label is gummed overall, because the picker arms slide over the whole label surface before the label leaves the magazine. Also, the placing arm incorporates a vacuum holder to assist in placing small labels in the correct positions on shoulders and other awkward places. In the 120 model (the number indicates the average speed per minute) each station is duplicated, and the bottles move two places at a time.

In most of the machines described the movement or rotation of the bottle is made use of in placing the label, but the application of top-straps over the closure and neck, usually for tax purposes, is a tricky operation in which the labelling machine has to do all the work. A quick-setting adhesive with a strong bond is needed. Fig. 11.8 shows the action of the World CS Stripstamper.

Figure 11.8. World CS Stripstamper applying top-straps

HEAT-SENSITIVE LABEL MACHINES

Machines for heat-sensitive labelling are basically similar to those described above, the essential difference being that the label is made sticky by placing it in contact with an electrically heated plate, to heat its adhesive surfaces to about 100°C. The 'Pony Label-Dri' semi-automatic machine for example is very similar to the wet-glue machine shown in Fig. 11.3, except that the label is moved in two stages, first to the heater plate and then to the bottle. The limitation of this method is that the paper shrinks as a result of dehydration when it is heated, and will expand slightly after sticking to the bottle, possibly causing blisters. This is insignificant for small labels, but can be troublesome if the label is wider than about 100 mm. Also the temperature puts some restrictions on the choice of varnishes, some of which soften at less than 100°C.

Automatic labelling is not very common, but Morgan Fairest have the 'Thermorfair' machine which is rated at 170 bottles per minute. Automatic machines have also been developed for handling heat-sensitive labels in continuous reels, with a photo-electric or similar device for cutting off the length required for each bottle. The New Jersey 'Label-Dri Champion' is an example, rated at 240 bottles per minute.

Figure 11.9.   Exodus Lorraine pressure sensitive labeller

PRESSURE-SENSITIVE LABEL MACHINES

Pre-cut pressure-sensitive labels can be obtained in sheet form for manual labelling; the individual labels have to be removed by hand from the backing sheet. Alternatively, they may be supplied in reel form, and simple mechanical dispensers are available.

Automatic machines require a supply in reel form; for example, the Exodus Lorraine, seen in Fig. 11.9, detaches body and foot labels from their backing and applies them to a range of Long John whisky bottles, giving throughputs of 180 per minute for the smallest (0·375 litre) size. The machine is fitted with a 'spotting' station to ensure that the labels are located into a decorative feature on the bottle.

Whitehall Machinery supply a semi-automatic Pony Placer, which is a pressure-sensitive version of the Pony Labelrite and provides outputs of up to 60 containers per minute. Their Autolabeler is a fully automatic machine operating at up to 300 bottles per minute.

Automatic machines are also available which cut and apply lengths from a continuous reel of pressure sensitive tape.

Lists of labelling machinery

Four lists are given:
1. Automatic machinery for labelling with wet glue,
2. Semi-automatic machinery for labelling with wet glue,
3. Equipment for applying heat-sensitive labels,
4. Equipment for applying pressure-sensitive labels.

Abbreviations generally refer to the positions on bottles to which labels can be applied:

B: Body.
S: Shoulder.
N: Neck.
T: Top-strap (over the top of the closure).
2: Applies labels to both sides of bottle simultaneously.
W: Wraps the ends of a long label all the way round.
R: Rolls a label right round a cylindrical container.
X: With spot register of label to bottle.

Names of machines are given where applicable; 'various' indicates that different names are used for different models in a range. Labelling speeds depend to a considerable extent on the complexity of the job being carried out, and it is not possible to quote simple comparative figures for different machines.

Automatic machinery for labelling with wet glue

Supplier	Round cross-sections only	Machine name and/or details	Round and other cross-sections
Adams Powell	R		
Anglo Continental			
(Hapa)	Universal		
(Weiss)	B	Various	BSN.2WX
Beth	R		
Bosch (Strunck)	Various BW	Various	BSNT.2
Bramigk (Simonazzi)	Irfas B	Various	BSNT.2WX
Crown (Baele Gangloff)		Various	BSNT.2WX
Erben (Anker)		Various	BSNT.2WX
(MA.CO)		Cometa	BW
(MA.CO)		Minor	BSN.2
(Seitz)		Various	BSN.2WX
(Weiss)		Various	BSN.2WX
Flower	Various BW		
Goddard (Kronseder)	Krones BSNT.2WRX	Various	BSNT.2WRX

Machine name and/or details

Supplier	Round cross-sections only		Round and other cross-sections	
Gravfil			Harley-Hinds	B.2X
Hart (Brüser)		R		
(Labelette)		R		
Jagenberg	Solar	BSN.2WX		
Mather & Platt (Standard-Knapp)	Various	BR.S		
Meyer			Various	BSNT.2WX
Morgan Fairest	Various	BSN.2WRX	Various	BSN.2X BN.2X BSNT
Newman	2VL/C	BR	2VL/CR	BRX
Precision		B.2RW		B.2RW
Purdy	Rotamatic	BSN	Linamatic	BSN.2X
	New Way	R	Unimatic	BN.W
Rockwell Pneumatic Scale (Pneumatic Scale)			McDonald Direc-Transfer	B.2X
Skerman			Various	BSN.R
Stackpole	Phin	R	Phin	BSNT.2WX
Whitehall Machinery	New Jersey	B.2W	New Jersey	B.2

Semi-automatic machinery for labelling with wet glue

Supplier	Machine name	Details
Bosch (Strunck)	Various	B
Erben (Anker)	Karat	B
	Junior	BS
	Standard	BS, and N or T.2W
Hart	Colabel–Sumbel	BS
Jagenberg	Various	B or S or N or T.W
Mayflower	—	BWR
Newman	B	R
	24A	BX
	Wet glue label dispenser for hand labelling	
Purdy	Universal	BS, and N or T.W
Stackpole	Phin	BT
Whitehall Machinery	Pony Labelrite	BSNT. WRX

Equipment for heat-sensitive labels

Supplier	Name of machine*	Details
Anglo-Continental (Weiss)	Jowe-4; A	BSN.W
Bosch (Strunck)	EHR A/B 01; A	—
Morgan Fairest	Thermorfair; A	—
R. A. Jones	Jones; A	B
Precision	Precision; A	BSN.2WX
Whitehall Machinery	Pony Label-dri; A	BSNT.2WR

* Abbreviations: S/A Semi-automatic; A Automatic.

Equipment for pressure-sensitive labels

Supplier	Name of machine*	Details
Avery	Various, S/A and A	—
Bosch (Strunck)	Various, S/A and A	—
Doran	Various, S/A and A	—
European Machine Systems	Exodus Lorraine, A	BSNT.2WRX
	Exodus Locator, S/A	—
Harlands of Hull	Combina, S/A and A	—
King	LH3	BSNT.R and 3 sides of rectangular bottle
Minnesota (3M)	Labeller, A	Uses unbacked tape
Newman	Facillette, A	—
	2VL/P, A	—
Precision	Various, S/A and A	—
Sellotape Products	Various, S/A and A	—
Sessions	Gusag Major, S/A	—
	Collomat A	—
Whitehall Machinery	Pony Placer, A	BSNT.2WR
	DiverSoMatic, A	BSNT.2WR
	Autolabeler, A	BSNT.2WR

* Abbreviations: S/A Semi-Automatic; A Automatic.

12. Packing, storage and transport

It is convenient to discuss the packing of bottles under three headings:

1. Individual packs, used for single bottles.
2. Multipacks, in which small groups of bottles are assembled and so presented at the point of sale.
3. Transit packs, i.e. all conventional crates, cases, etc., used for large groups of bottles, plus the newer wrap-around cases and shrink-wrapped assemblies.

INDIVIDUAL BOTTLE PACKS

The packing of bottles in individual cartons, boxes, protective sleeves, etc., is mainly confined to products which can bear the additional cost, such as cosmetics, some pharmaceutical products, and wines and spirits.

Individual packing of very large bottles is probably best done by hand, but machinery is available for the automatic cartoning of medium and small bottles. For example, Rose Forgrove make four different cartoning machines operating at maximum speeds of 60 to 200 bottles per minute.

MULTIPACKS

The multipacking of bottles, usually in groups of 3 to 12, is now a popular method of merchandising. Multipacks not only encourage customers to buy more than one bottle at a time; they also assist in handling and display in the retail store. They are not, however, intended to be sufficient in themselves for bulk handling and transport.

One simple method which has been used successfully is to enclose the group of bottles in a band of moistened cellulose film, which locks them firmly together when it shrinks. A similar band could be made with adhesive cellulose tape, provided that it did not disfigure the label when removed.

Of greater general interest are the various carton-board 'carry-home' packs now used widely for beer and mineral bottles. These are broadly of two main types. The first consists of a simple sleeve wrapped round the group of four or six bottles, with cut-out portions and raised tabs to hold the bottles in position, and a cut-out hole in the top for carrying. Among the most popular packs of this type are the Cluster-Pak, available from DRG Multiple Packaging and Mardon, Son & Hall, and the Jak-et-Pak made by Metal Box. All three companies can supply a range of multipacking machinery. Jak-et-Pak equipment, for example, extends from simple hand jigs to fully automatic machinery catering for outputs from 6 to 150 packs per minute. Cluster-Pak machinery will provide outputs ranging up to 250 or 300 per minute. Some typical Cluster-Paks, supplied by DRG Multiple Packaging, are shown in Fig. 12.1.

Figure 12.1. Examples of individual packs, multipacks and transit packs

Another type of multipack consists of a narrow square-section tube of carton board with die-cut holes in the top and bottom walls, designed to clip over the necks of the bottles. Compared with the full wrap-round sleeve, this has the advantage of using less board and is therefore cheaper. It does not offer the same opportunities for promotional printing as the full sleeve because of its smaller display area, but on the other hand it does leave the labels and bottles on uninterrupted view, which can be turned to marketing advantage. The main examples are the Link Pack, supplied by Bowater Packaging, and the Carrybar, supplied by Mardon, Son & Hall.

Machinery is available for automatically locking the clips over groups of bottles giving outputs in the case of the Link Pack system of 120 three-bottle packs per minute. Two-bottle and four-bottle Link Packs can be handled at a corresponding speed (i.e. 360 bottles per minute). Bottles are fed to the machine by a conveyor, direct from the filling, capping and labelling line.

Another multipacking system, called the Hi-Cone, has been designed primarily for multipacking cans in a polythene film carrier. The system, available in the UK from ITW, is now being developed for multipacking carbonated soft drinks in bottles and could prove a rival to the carton-board multipacks in the future.

Finally, groups of bottles can be shrink-wrapped together, usually in a shallow tray; this system is described on page 280.

TRANSIT PACKS

The essential functions of transit packs are to facilitate efficient storage and transport, and to ensure that the bottles reach their point of sale clean and undamaged. Reusable packs are generally made of wood, metal or rigid plastics. Until recently, non-returnable ones were almost always made of fibreboard, but today shrink-wrapping in plastics film is becoming more and more popular for distribution within the UK. The reasons for this are discussed later.

Nearly all returnable bottles are packed in returnable cases or crates, for the very good reason that this is by far the cheapest way. The designs of modern returnable cases have been well worked out, and little comment on them is needed here. They must, of course, have a shape and size suitable for stacking (this point is discussed in greater detail for fibreboard cases, on p. 272).

Special attention has been given to making the case divisions suitable for automatic loading and unloading of bottles, as may be seen with the plastic separators in wooden beer cases. For returnable mineral bottles, shallow wooden trays can be used. These require a little more care in handling and stacking than do deep cases, as the whole weight of a stack of filled trays has to be carried by the bottles and closures. This they can do easily if the weight is spread evenly, but if the ground is uneven or the stacking careless the load distribution may become very unequal. Shallow trays lend themselves very well to automatic loading and unloading.

If returnable crates without divisions are used, this must be taken into account when designing the bottle, to ensure that it will be strong enough to withstand the rougher treatment which it will receive as a result. On the other hand, if divisions are used, it is important to see that they really do divide the bottles, keeping them apart even when the crate is shaken or impacted violently.

The main requirement of rigid cases and crates is that they should protect bottles against direct impacts. Little cushioning effect can be provided so the bottles must be strong enough to withstand the violent deceleration or acceleration caused when a case is dropped or struck. Fibreboard cartons and trays, on the other hand, usually combine some impact protection with some cushioning, and the secret of good design is to get sufficient of both kinds of protection, at the lowest cost.

Let us now take a closer look, firstly at fibreboard cases and secondly at shrink packs.

Fibreboard cases

Before going into detail about the design of fibreboard cases, it is necessary to review briefly the different materials used for making corrugated fibreboard, the different types of board which can be constructed, and the technical terms used to distinguish them. The nomenclature is unfortunately very confusing, and the packer may understandably be tempted to leave the whole subject for his glass supplier and case maker to sort out. For those who want to know all the answers, however, the following notes may be helpful.

MATERIALS USED FOR MAKING FIBREBOARD

Kraft paper or kraft liner: A tough and smooth paper, made by a chemical process from long wood fibres, usually Scandinavian spruce or pine. It is naturally brown, but can be bleached to near white, or dyed.

Semi-chemical paper: This is widely used for corrugated fluting. It is made from short wood fibres by a process which is partly chemical and partly mechanical.

Chipboard: A paper made wholly or mainly from waste paper.

Test liner: A paper in which the body of the paper is chipboard or a similar cheap material, with a thin surface lamination of kraft paper. (The reason for the name is historical, relating to the time when it was necessary to test each production batch.)

TYPES OF FIBREBOARD

Solid: A homogeneous or laminated board, usually lined on one or both sides with kraft or similar smooth paper. Used mainly for trays, and also for divisions or partitions in corrugated cases.

Corrugated – single-walled.   A sheet of corrugated paper sandwiched between two flat sheets.

Corrugated – double-walled.   A double-decker sandwich, with two corrugated sheets and three flat sheets.

Corrugated – triple-walled.   A triple-decker sandwich with three corrugated sheets and four flat sheets. Used for very heavy duties, and especially for export consignments.

DIMENSIONS, WEIGHT AND STRENGTH

The main guide to the serviceability of a board is its weight and thickness, but the 'bursting strength' and resistance to water penetration can also be measured directly.

Caliper.   This is a name given to the thickness of a plain flat board or paper, measured in millimetres.

Board weight or 'substance'.   The weight, in grams per square metre, of a flat board or paper. The usual weights are 112, 125, 150, 175, 200, 225, 250 and 300 g/m².*

Fluting.   There are four sizes of fluting:

	Corrugations per metre	Height of corrugations
A flute	105–125	4·5–4·7 mm
B flute	150–185	2·1–2·9 mm
C flute	120–145	3·5–3·7 mm
E flute	290–320	1·1–1·2 mm

Note that C is intermediate between A and B. C and B are the most commonly used, and in double-walled board it is common to use one layer of C flute and the other of B flute.

* Sometimes incorrectly abbreviated as 'gsm'.

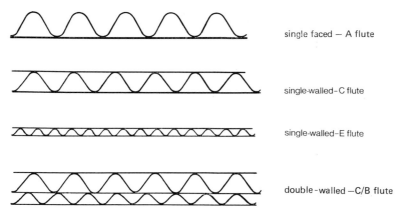

single faced — A flute

single-walled-C flute

single-walled-E flute

double-walled —C/B flute

Figure 12.2.   Types of corrugated fibreboard (shown 50% over full size)

SELECTION OF MATERIALS FOR CASE CONSTRUCTION

Kraft is the toughest and smoothest paper, and it also has some resistance to water, so this is the best material for the outer facing and the inner backing for corrugated board. For a cheaper board the kraft may be replaced by test liner or even chipboard. The best fluting is made from semi-chemical paper, but in the cheapest boards ordinary chipboard may be used. Similar materials are used for double-walled board.

The board weight used depends on the pack construction, weight of contents and likely transit hazards. Export packs, for example, will need to be stronger than home-trade packs.

Case construction and board weight may also depend on whether the pack is to be carried at 'owner's risk' or 'carrier's risk', a matter discussed more fully later in this chapter.

Finally, the appearance and decoration of the packs should be considered. Various coloured, bleached or dyed boards are available (at extra cost) if required, and they can help to make a distinctive pack. If, however, the packs have to undergo a lot of handling, the inevitable dirty marks and smudged ink will show up more noticeably on a light-coloured board than on the natural brown board, and the result may not be as good as expected.

DESIGN OF FIBREBOARD CASES AND TRAYS

A successful pack provides four things:

1. A means of handling and transporting the bottles efficiently and economically;

2. A means of storing the filled bottles, and if necessary the empty bottles before air cleaning, in good condition;

3. Protection against impacts received in normal handling and transport;

4. Protection against indirect stresses, such as the internal pressure effect when a liquid-filled bottle is suddenly decelerated.

It is not possible to lay down a complete set of rules for designing a pack, but at least we can consider how these four requirements affect the decisions to be taken.

HANDLING AND TRANSPORT REQUIREMENTS

Ease of handling and transport is affected considerably by the shape of the pack. Building a stable stack of cases, whether on pallets or on the ground, is very much like bricklaying, and the proportions of the standard brick have been chosen to provide the maximum number of different ways of interlocking different rows and courses of bricks so as to obtain the strongest structure. In a stack of fibreboard cases, frictional forces between the layers serve exactly the same purpose as the cement in a brick column, and a 'brick shaped' case would be very satisfactory. This ideal shape cannot often be achieved exactly, as so much depends on the number and shape of bottles to be included, but every attempt should be made to avoid a cube shape, or any pack with a square base. The 4×3 arrangement, which is usual for a pack of a dozen bottles, is not really ideal; 5×3, 6×3 or 6×4 would be better. And whenever possible the case's horizontal dimensions should be chosen to be exact subdivisions of pallet dimensions, so they can be stacked without wasting space.

As mentioned earlier, the type and strength of the case should be chosen to suit the method of transport; for example, road transport in covered vehicles may make a simple open tray possible. Similarly, the methods to be used for packing and unpacking the case may affect the design.

It should be stressed that a packer is quite free to use any type or strength of pack he wishes. If a carrier regards a pack as 'sub-standard' he will not usually refuse to handle it, but he may handle it at 'owner's risk' rather than 'carrier's risk'. (Sometimes this turns out to be much cheaper for the packer than using an approved pack. Reputable fibreboard case manufacturers are always willing to offer suitable case designs.)

A normal method of obtaining the approval of carriers or insurance companies for a pack is to make it in accordance with the standards laid down by the Fibreboard Packing Case Manufacturers' Association

(FPCMA). If this is done a certificate will be granted, to be printed on every individual pack. This is generally accepted as evidence of sufficient strength. There are separate standards for carriage within the UK mainland and for carriage overseas.

The main area in which certified cases are demanded is for export whisky packs. A typical FPCMA double-walled case for up to 20 kg of wines and spirits in bottle would consist of inner and outer kraft or test liners with a combined weight of 450 g/m², and a middle chipboard wall of 125 g/m² and fluting with a minimum weight of 112 g/m². The fluting must comply with stipulated flat crush tests and the liners with specified water resistance tests. Double-wall partitions are required between the bottles.

To take another example, a certified single-walled open-topped full depth case to carry up to 15 kg of non-alcoholic goods (e.g. jam) within the UK mainland must have inner and outer liners adding up to 450 g/m² and a fluting of at least 112 g/m². Partitions must be provided consisting of chipboard inner and outer facings with a combined substance of at least 350 g/m². Alternatively, kraft or test liners can be used with a combined substance of at least 300 g/m².

To qualify for FPCMA certification, cases must not only be made of the correct substance and type of board, but must also conform structurally to certain case styles selected from the International Case Code.

For export to the USA, packers may need to comply with the rather different packaging regulations of the US railroads. These regulations also apply to road transport in the USA and, with minor variations, to the Canadian railways. These standards tend to be on the high side; in fact, the US standards for domestic packaging are very similar to the FPCMA standards for export packaging. The basic difference is in the substance of the fluting required, since the US regulations insist on a minimum of 127 g/m², whereas our export standards call for a minimum of 112 g/m².

STORAGE REQUIREMENTS

The first question is whether the filled bottles need to be retained indefinitely in dust-free condition; if so, a closed case rather than an open tray is essential, unless the entire pallet load is to be covered. Also, cut-out hand-holds should not be used. The same applies if the same cases are to be used for supplying empty bottles to an air-cleaning machine. Much of the dust which an air cleaner has to remove from the bottles comes from the edges of the fibreboard itself; this can be minimised by using a good quality board, and ensuring that the case maker cuts the board cleanly, without leaving ragged edges.

If tall stacks of cases are left in store for considerable periods they may occasionally show a tendency to sag or tilt. This may be due to inaccurate stacking in the first place, but another reason may be inadequate rigidity of the cases themselves. Obviously the strength of board should be chosen bearing in mind the likely height and weight of the stack. The best method of preventing sagging is to arrange that the height of the capped bottles takes up very nearly all the available height inside the case, so that if the case starts to sag it will be kept in shape by the bottles themselves. In other words, in a good stable stack, the bottles support the cases, not vice versa.

IMPACT PROTECTION

Protection against impacts is obtained either by providing a rigid shield to ward off impacts (like the armour plate on a tank) or a yielding barrier to absorb the energy of the impacting object (like a parapet of sandbags).

In the main, fibreboard uses the second of these methods, as the rigidity of double-walled board is not very high. When a corrugated liner is fitted inside a case, with its corrugations at right angles to those in the wall itself, the rigidity is considerably improved.

If it is certain that impact resistance is much more important than cushioning, good results may be obtained by using solid board instead of corrugated.

It is not possible to generalise about what level of impact a case should be able to withstand; indeed it is by no means easy even for an expert case designer to decide exactly what will be needed in practice. It is easier to err on the side of caution if the price of the product will stand the extra packaging cost, but in order to determine what is really the most economic level of packing there is no substitute for carefully conducted trials under service conditions. Bottle makers and case makers have acquired a great deal of experience in this way, and their advice may save the packer unnecessary expense.

CUSHIONING

When a moving bottle is suddenly decelerated, as, for example, when a falling case hits the ground, the forces on the bottle will cause stresses in the glass. If the bottle is filled with liquid, the momentum of the moving liquid will exert a transient internal pressure. In both cases, the magnitude of the stresses will depend on the rate of deceleration; the purpose of cushioning is to restrict the deceleration, and so prevent dangerous stresses. It is possible to provide almost any amount of cushioning: there are some bottle packs, containing liquids, which can be dropped from 2 to 3 metres without damage. Generally, however, the cost does not justify such measures, particularly because the chance of such violent drops occurring in practice is remote. Drops of up to 0·5 metre are not uncommon in manual handling of cases (but are to be avoided if at all possible), and in many situations it has been found that a liquid pack designed to survive a drop of 0·6 metre represents the economic limit of cushioning. (One of the advantages of the use of pallets is that it considerably reduces the chance of such drops occurring.)

Cushioning is at present provided most effectively by corrugated board, in the form of layer pads above and below the bottles, liners round the sides (already mentioned in connection with impact protection), and divisions between the bottles. In recent years it has often been possible to reduce the amount of cushioning and the conditions of handling have become better controlled, so divisions are often made successfully and economically from thin solid board. Their main function then is to prevent glass-to-glass contact, but they do also provide a little cushioning.

Corrugated fibreboard is not a perfect cushioning material because it has limited powers of recovery after being compressed or struck; but there are no better materials available at comparable cost.

CASE ASSEMBLY

Most conventional cases are made from one pre-cut piece of board and are supplied flat, with one seam made by the case maker. Assembly therefore simply consists of opening out the folded case, and gluing or otherwise fixing the bottom flaps into position. This is still often done manually although automatic case erectors are becoming increasingly popular.

If the cases are to be used more than once, the bottom flaps will usually be fixed with gummed tape, but if the case is a single-trip one, the choice of fixing is between glued flaps and wire stitching. Gluing generally gives the strongest and least pilferable joint, but stitching is sometimes used for heavy cases.

Bottle divisions are supplied ready interlocked, and only require opening out and inserting in the cases. Fig. 12.3 shows a typical double-walled export pack, supplied by Thames Case, with Lokfast divisions made from solid fibreboard.

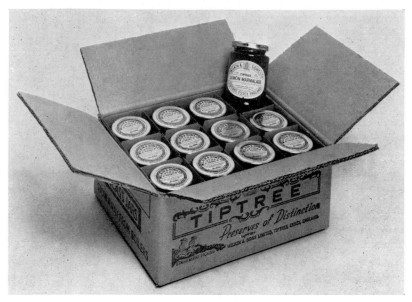

Figure 12.3.   Transit case with solid board divisions

Apart from the normal type of fibreboard case, there is increasing interest in the new type of pack known as the 'wrap-around' case. As its name implies, this consists simply of a cut-and-creased corrugated board blank which is automatically wrapped around the products being packed. Methods of packing both conventional and wrap-around cases are discussed below.

Methods of packing and sealing cases

Filling cases by hand has now become the exception rather than the rule, and at production speeds of over 100 bottles per minute, mechanical packaging is almost essential. Automatic packing will not be successful unless due attention is given to maintaining accurate case dimensions, and accuracy and squareness of case assembly, including, of course, the internal divisions.

Dual-purpose machines suitable for both unpacking and re-packing have already been discussed in Chapter 7. One machine of this kind is seen in Fig. 12.4. In addition there are a number of 'drop packers' designed solely for packing bottles into cases. Most of them work in more or less the same way. The bottles are led from a conveyor belt on to a marshalling table, where the required number of bottles is arranged in a squad of the right shape. Most machines do this simply by carrying the bottles forward

Figure 12.4. Kettner automatic case packer

on a broad moving belt into slots, with an overhead oscillating mechanism to spread the bottles evenly into all the slots. This method of marshalling works only with cylindrical bottles; but the British Miller Hydro and other case packers have a more complex motion for other shapes of bottles. When the squad is assembled, trap doors open and the bottles slide down into the waiting case, guided by spring steel or plastic fingers.

With small jars fitted with tinplate caps, it is possible to use an electromagnetic head to lift the assembled squad and lower them into the case.

The Certus range of drop packers can handle up to 20 cases per minute on a single-head machine and 30 cases per minute on a two-head machine. Units for even higher speeds can be supplied with a two-lane case conveyor and the product infeed can be adapted to handle square containers. Special versions are available for packing multipacked bottles into fibreboard cases.

Yet another method of loading bottles into crates, fibreboard cases or trays is provided by the Huntingdon Flowpacker, which gives outputs of more than 60 cases per minute. Bottles are fed into the machine and grouped into patterns on a continuous basis. The groups of bottles are then lifted by rubber spring-loaded grippers carried on an overhead chain conveyor. Cases are carried on a continuously moving escalator system which firmly grips each case and raises it around the group of bottles while they are still being conveyed by the overhead grippers. The case continues upwards until it supports the bottom of the bottles, when the grippers relax their hold. Cases and bottles flow together continuously throughout the operation.

After the cases have been filled with bottles, and the top layer pad placed in position, if required, the assembly is completed by fixing the top flaps. Several types of automatic case sealer are available for gluing the flaps and/or fixing them with gummed tape. It is to be noted that cases for dutiable liquor are usually required to have their seals completed with gummed tape, as a witness against pilfering.

'DROP-AROUND' CASE PACKER

In the Hermier drop-around case packer the operations of case erecting, filling and sealing are all combined in a single machine. This differs from an ordinary case packer in that the case is dropped over a batch of bottles, instead of the bottles being loaded into the case. First of all, a collating and selector system arranges the bottles in batches. Then a flat case is squared and dropped over the batch, leaving all flaps open. (This operation can be done either manually or automatically.) At subsequent stations, the top and bottom inner flaps are closed around the batch and the cases are driven steadily past the gluing heads. The outer flaps are then automatically closed, and the glued cases pass through a compression section, to allow the glue to set, before ejection on to a discharge conveyor.

'WRAP-AROUND' PACKERS

The wrap-around technique, whereby a flat corrugated board blank is wrapped round a group of products, has been successfully used for bottled goods in several countries and is likely to become of increasing importance in the UK. Like the 'drop-around' packer, mentioned above, its main advantage is that the three separate operations of case erection, filling and sealing are all combined in a single machine. In addition, the wrap-around blank uses much less board and is easier to produce than a conventional case. It is therefore cheaper to buy and takes up less space in storage.

Fully automatic wrap-around machines are available which collate the bottles into the required groupings, wrap the corrugated blank around them to form the case, and seal the pack, all in a continuous sequence of operations. Semi-automatic wrap-around machines which rely on manual loading of blanks and bottles are also available.

The Pak-Master system is available in a number of versions for different products, pack sizes and outputs. The most adaptable machine in the range is the Pak-Master III, which is built for relatively quick change-overs. Flat blanks are stacked in a hopper and fed one at a time on to the loading station. Meanwhile, bottles are collated in an in-feed section designed to suit each individual product. A group of bottles is placed on the blank, which is then automatically wrapped around them while a combination of cold glue and hot-melt adhesive is applied to the flaps (or hot-melt only, if required). The case then enters a compression section, where it is squared completely before the side flaps are closed by means of pneumatically controlled pressure plates. Outputs of up to 17 cases per minute can be obtained on the Pak-Master III, but other versions are available to give outputs up to 25 cases per minute. Fig. 12.5 shows a Pak-Master machine packing half-bottles of Teacher's Whisky.

Figure 12.5.   Pak-Master wrap-around case packer

Other wrap-around packers include the Emhart EZ–1259, which gives outputs of up to 25 cases per minute, and the OCME series AV with outputs of up to 24 cases per minute. Besides forming a tight wrap-around case, some machines can also insert dividers or partitions automatically. The partitions are made on the machine from specially shaped interlocking pieces of fibreboard which it inserts between the rows of bottles.

By packing bottles very tightly on a wrap-around machine, it has been found possible to eliminate the need for internal partitions in some instances. This principle has been extended still further in the Pak-Master system by the use of a special taper-sided case, which exerts pressure on the shoulders of the bottles, giving a further insurance against rotation of the bottle in transit. This also provides label protection for glass containers.

The Focke & Pfuhl FPS 463 machine offers a novel approach to wrap-around packaging by using a new kind of case blank in which the internal partitions are punched out from the blank itself by the machine. Bottles are fed to the machine in one or more rows, then formed into two rows. Next, they are arranged in groups of 6, 8, 10 or 12 bottles. Blanks are taken from a magazine and hot-melt adhesive is applied to the side flaps, which are then folded up. The partly formed case is wrapped around the group of bottles and the internal partitions are erected between them. The case leaves the machine sideways, and is then completed by the gluing and closing of the outer flaps. Outputs can be as high as 80 to 100 cases per minute.

Shrink-wrapped packs

Plastics films which have been stretched or 'oriented' during manufacture will shrink again when they are heated. This principle is exploited to very good effect in the various forms of shrink-wrapping now widely used for many kinds of consumer products. Shrinkable films can be made from a wide variety of thermoplastic materials, but for reasons of strength and economy, oriented polyethylene is the most popular film for shrink-wrapping such products as bottles, cans and cartons.

For some purposes, acceptable shrink packs can be produced simply by wrapping and sealing a group of bottles in oriented polyethylene film and passing the wrapped bundle on a conveyor through a heated shrink tunnel. At present, however, it is more usual to collate the groups of bottles on trays made from solid or corrugated board before wrapping them, since the shrink energy of the film is not sufficient to pull heavy bottles together to make a tight wrap.

The trays provide a rigid base which makes the bundles easier to stack and handle. They are invariably supplied to the packer in the form of creased and slotted flat blanks which can be erected into trays either manually or by machine. Trays for manual assembly are normally of the self-locking type, in which the corner flaps are secured by a simple arrangement of tabs or folds. Machine-erected trays can be of similar construction, or secured by automatic corner-gluing with PVA or hot-melt adhesive. A corner-glued tray which is machine-erected on equipment supplied by Bowater Packaging, is seen in Fig. 12.1.

For high-speed lines, equipment is available to collate and load the bottles into the trays automatically. (In brewery and soft drinks plants, the ordinary crating machines can often be used to load trays.) The loaded trays are then passed to a sleeve wrapper, which loosely wraps them in an open-ended sleeve of shrinkable film. Next, the wrapped trays pass on a conveyor through a heated tunnel, where the films shrinks to form a tight, neat transit package.

Several types of shrink-wrapping machinery are available to apply the film, ranging from simple manually operated sealers to fully automatic machines. Some of these can be combined with automatic tray-loading equipment and shrink tunnels so that the operation of collating bottles, loading trays, applying shrinkable film and shrinking it round the loaded trays can be carried out in a fully automatic sequence of operations.

In the Micronerrani shrink-wrapping system, groups of bottles can be shrink-wrapped without the need for supporting trays, although for heavy loads (and ease of stacking) a flat sheet of board can be used. Products are conveyed to a loading module which collates them into the required group pattern. If a supporting sheet of board is needed, this is automatically fed from a magazine and the bottles are collated on top of it. The entire load is then pushed automatically against a vertical curtain of shrinkable film formed from two reels of film, one above and one below the working station. As the load moves forward, the film is drawn round it and a wide seal is made at the back of the pack to form an open-sided sleeve. A photocell controls the descent of the sealing bar. Finally, a belt conveyor carries the pack through a shrink tunnel.

Finished shrink-wrapped packs are usually palletised (automatically or by hand) for storage, and sometimes also for delivery to wholesalers or large retailers. Most of the major retailers have welcomed the growth of shrink-wrapping because the film and trays employed are easier to dispose of than large quantities of fibreboard cases. Removing the film and unloading the trays is also a simpler, quicker operation than opening and unloading cases. In many shops, the trays are not even unloaded, but simply stacked on shelves or used to make floor displays.

From the packer's point of view, shrink-wrapping enables the cost of transit packing to be reduced, though the full economic benefits of the system are only obtained if incoming glass is bulk-palletised, thus eliminating the use of fibreboard cases at the input end as well as the output end of the filling plant (see Chapter 7).

Shrink-wrapping cannot, of course, provide the same degree of impact resistance and cushioning that is offered by a fibreboard case, but provided they are handled reasonably, shrink-wrapped packs are able to provide satisfactory protection during transport. The fact that the contents can be seen through the film tends to encourage careful handling.

Handling and storing transit packs

PALLETS

Wooden pallets are commonly used for handling and warehousing filled bottle packs. The usual method of building up a pallet load manually is to arrange one layer at a time, and to build up sufficient layers to make a stack which is roughly cubic in shape. Accuracy of placing is very important.

Automatic pallet loaders are available which perform the same actions as in manual loading, even to the extent of arranging successive layers with the length of the cases or shrink packs running in different directions, to provide good interlock between the layers (Fig. 12.6).

Figure 12.6.   Alvey series 300 pallet loader, handling cases of Johnnie Walker whisky

Returnable plastic or metal crates are usually designed so that one crate locks into position on top of another, so for these a different method of stacking is needed. Automatic crate stackers make a vertical column of, say, five crates, which can then be handled individually on a hand truck, or formed up into a bulk load on a pallet.

In addition to flat wooden pallets, the use of metal cage pallets for delivering filled goods is being increasingly sought by operators of large supermarkets. These are stackable rectangular containers made from heavy-gauge wire, and usually have one side demountable so that consumers can serve themselves directly. The system lends itself quite well to bottles in multipacks, but if individual bottles are merchandised in this way they need to be stacked upright, with separating boards between the layers.

FORK TRUCKS

The use of pallets means the use of fork-lift trucks, of which there are many types and sizes available. Four general points concerning them will be mentioned here:

1. It is important to decide, first, the sizes of cases and pallets, etc., and then to choose the fork-truck. (In other words, this is one occasion when it is important to put the case before the horse.) The standard size of pallet used by the glass manufacturers, 1200 × 1000 mm, is also rightly chosen as standard by most packers.

2. The job of a fork-truck is to *lift*, and it is wasteful if it has to spend a large proportion of its time transporting packs over long distances on the level. Less expensive trucks are available simply for horizontal movements.

3. Careful supervision and training of truck drivers is necessary if accidents are to be avoided. The Royal Society for the Prevention of Accidents can provide useful advice on this subject.

4. The movement of pallet loads by truck is an operation which more and more companies are finding it profitable to automate. Systems for using both robot fork-truck and towing tractors are available commercially.

STORAGE

Bottle stores do not need to be elaborate in any way; the only essentials are a firm level floor, unimpeded as far as possible by pillars or stanchions, and plenty of height. Pallet loads of bottles can with care be stacked to heights of 10 metres or more.

The two enemies of good storage are dirt and water. Airborne dust or grit is the most troublesome, and if a warehouse is free from holes in the roof and walls, dirt can only penetrate in quantity when there are two open doors to permit a through current of air. The solution is obvious. It is, however, unrealistic to expect a fork-truck driver to keep getting off his vehicle to open and shut doors; there are several simple ways of arranging for doors to open and close automatically, or to have them controlled by switches which a driver can reach without dismounting.

With regard to water, it should be easy to keep the rain away with good gutters and drains, etc., but it is just as important to discourage the water which will condense on bottles and cases if the temperature and humidity of the air vary too much. (Air at any temperature will contain a certain amount of dissolved water vapour, and the relative humidity is the amount of dissolved water, expressed as a percentage of the maximum possible amount which the air can retain at that temperature. As air is cooled, its maximum possible water content decreases, and so the relative humidity increases, eventually to 100%. If further cooling occurs, some of the water vapour is bound to appear as condensed liquid.)

If fibreboard becomes wet it soon looses its strength and resilience, and will not be able to provide proper protection for the bottles. If empty bottles become alternately wet and dry during storage, air cleaning will be more difficult because the dust will tend to stick to the glass. Varying atmospheric conditions can eventually affect the brightness of the glass surface itself, and washing may be necessary to restore the original appearance.

Direct control of the humidity in a warehouse is not an economic proposition, but sufficient control may usually be obtained simply by keeping an eye on the temperature. (Extremes of temperature are likely to be bad for the product, anyway, and freezing weather can in some cases upset label adhesion.)

Sudden temperature changes will be avoided by keeping the doors shut, as mentioned above in connection with cleanliness. A little wall or roof insulation, particularly on the side of the prevailing wind, may also be worth while. Only in extreme cases has it been found necessary to provide any artificial space-heating.

The final point about storage is the need to use stocks in rotation. This applies obviously to filled bottles, but it is also desirable for stocks of new empty bottles.

Lists of packing machinery

Bottle cartoners – individual packs

Supplier	Machine name	Details*	Max speed per min.
Arenco	Arencopac 1000	Inserts leaflet, A	120 cartons
	Arencopac 2000	Inserts leaflet, A	200 cartons
Anglo Continental (Zanasi Nigris)	KA	Folds and inserts leaflet, continuous motion, A	270 cartons
	KB	Folds and inserts leaflet, intermittent motion, A	100 cartons
Bosch (Robert Bosch, Hoefliger & Karg)	Various	Intermittent motion, Continuous motion, Leaflet insertion, Carton coding and reading Bottle in-feed, S/A or A	400 cartons
Freezepack	IM5	Hand operated	70 cartons
R. A. Jones	IMV	Inserts leaflet, codes, A	60 cartons
	CMV	Inserts leaflet, codes, A	120 cartons
	CMC-200	Inserts leaflet, codes, A	200 cartons
	CMC-400	Inserts leaflet, codes, A	600 cartons
	CMH	Inserts leaflet, codes, A	180 cartons
Kane	Kanepack	Hand or auto insertion	60 cartons

* Abbreviations: S/A Semi-automatic; A Automatic.

Bottle cartoners – individual packs (continued)

Supplier	Machine name	Details*	Max. speed per min.
Pacunion (IWKA)	Cartopac (Various)	Inserts leaflet, A	40 to 400 cartons
Rockwell (Restelli)	Arvella	Intermittent motion, A	50 cartons
	ARV	Continuous motion, A	160 cartons
Rose Forgrove	RF390	Inserts leaflet, codes, hand or auto insertion	60 cartons
	RF26	Inserts leaflet, codes, A	100 cartons
	RF80	Inserts leaflet, codes, hand or auto insertion	120 cartons
	RF21	Inserts leaflet, codes, A	200 cartons
Whitehall Machinery	Vertuck 74	Inserts leaflets, codes, A	120 cartons
	Vertuck 72	Inserts leaflet, codes, A	75 cartons
	54L/82	Codes, A	60 cartons

* Abbreviations: S/A Semi-automatic; A Automatic.

Multipackers ('Carry-home' packs, etc.)

Supplier	Machine name	Details*	Max. speed per min.
Bowater	Link Pack	Grips bottles by necks, A	120 packs
	Certipack	Wraps sleeve round bottles, A	250 packs
DRG Multiple Packaging	Cluster-Pak	Wraps sleeve round bottles, S/A, A or manual	250/300 packs
	Cluster Tray	Cluster-Pak sleeve with integral tray (shrink facility not required)	—
Holstein & Kappert	—	Wraps carton round bottles, A or manual	85 packs
R. A. Jones	Various	Erect cartons, collate and insert 12 bottles	150 packs
Mardon	Cluster-Pak	Wraps sleeve round bottles, S/A and A, or manual	250 packs
Metal Box	Carrybar	Grips bottles by necks, A	100 packs
	Jak-et-Pak	Wraps sleeve round bottles, S/A and A, or manual	150 packs

* Abbreviations: S/A Semi-automatic; A Automatic.

Tray erectors (for shrink-wrapping)

Supplier	Machine name	Details*	Max. speed per min.
Amdelco (Systems)	Various	Glued trays, A	25 trays
		Self-locking trays, S/A and A	40 trays
Andrew & Suter (ABC)	Traymaker	Glued trays, using hotmelt adhesive, A	30 trays
Bosch (Robert Bosch–Hamac Hoeller)	TFB	Glued trays, using hotmelt	30 trays
Bowater	CFG	Glued trays, PVA or hotmelt, S/A or A	60 trays
Crown (Huntingdon)	HEO	Glued trays, A	75 trays
Emhart (Standard-Knapp)	Roto–Tray	Glued trays, A	60 trays
Engelmann & Buckham (Pont-a-Mousson)	Seva 03	Glued trays, A	25 trays
Engineering Developments	Various	Glued trays, A	25 trays
		Self-locking trays, S/A and A	40 trays
Erben (De Hoop)	Various	Self-locking or glued trays, S/A or A	50 trays

Supplier	Machine name	Details*	Max. speed per min.
F. & D. Packaging	BL.1	Glued trays, A	120 trays
Freeze Pack	Free-Lok	Glued, hotmelt or self-locking Hot-air double-wall fold end lock	120 trays
Inpac	Bell	Glued trays (cold glue or hotmelt), S/A or A	60 trays
Kane	Kanepack/TLE	Glued trays using hotmelt or pressure sensitive adhesive	25 trays
Mardon	—	Glued trays, S/A, A or manual	50 trays
Metal Box	Kolatarap	Self-locking trays, S/A and A	40 trays
Reed	Reed-Kisters	Glued trays, A	50 trays
Rockwell/Iwema	—	Glued trays, A Self-locking trays, A	60 trays 40 trays
Thames Case	Kolataform EB	Glued trays, S/A	12 trays

* Abbreviations: S/A Semi-automatic; A Automatic.

T

Shrink-wrappers

Supplier	Machine name	Details*	Max. speed per min.
Amdelco (Systems)	Various	Sleeve wrappers, S/A and A Tray erector/loader/wrapper, S/A and A	40 packs
Borden	Various	Sleeve-wrappers, S/A	15 packs
Bosch (Robert Bosch–Hamac Hoeller)	Various	Tray-wrapper Sleeve-wrapper, S/A and A	35 packs
Breda	Various	Sleeve-wrappers, A	20 packs
Crown (Huntingdon)	Wrapcap	Tray-wrapper, A	70 packs
Engineering Developments	Various	Sleeve-wrappers, S/A and A Tray erector/loader/wrappers	40 packs
Erben (De Hoop)	Various	Sleeve-wrappers, S/A and A	32 packs
Freeze Pack	Pester	Stretch- or shrink-wrappers	20 packs
Hänsel	—	Sleeve-wrappers, A	25 packs
I. D. Packaging	MSS	Sleeve-wrapper, S/A	8 packs
	RSM	Sleeve-wrapper, A	15 packs
	RSQ	Sleeve-wrapper, A	25 packs

Supplier	Machine name	Details*	Max. speed per min.
Inpac	CM	Continuous-motion sleeve-wrapper, A	75 packs
	ILA	Sleeve-wrapper, A	30 packs
	SW/XF	Sleeve-wrapper, A	30 packs
	SSA	Sleeve-wrapper, S/A	14 packs
	HSW	Sleeve-wrapper, foot operated	10 packs
	ISA	Sleeve-wrapper, A	15 packs
Kane	Kanepack/Unity	Tray loaders and sleeve-wrappers, S/A and A	30 packs
Lerner	Various	Sleeve-wrapper, A	30 packs
		Full overwrapper, A	20 packs
Mather & Platt	PSW	Sleeve-wrapper, A	14 packs
	RSW and RSW Twinwrap	Sleeve-wrappers, A	60 packs
(Huntingdon)	Wrapcap	Tray-wrapper, A	70 packs
Reed	Reed-Kisters Type 90	Sleeve-wrapper, A	60 packs
	Reed-Kisters Type 44	Tray-erector, loader-and-wrapper, A	25 packs
	Reed-Kallfass	Sleeve-wrappers, S/A, A	60 packs

* Abbreviations: S/A Semi-automatic; A Automatic.

Shrink-wrappers (continued)

Supplier	Machine name	Details*	Max. speed per min.
Rockwell/Iwema	—	A	80 packs
Rose Forgrove	UE6	S/A and A. Interchangeable for sleeve-wrapping, full overwrapping, or tray wrapping	40 packs
Simon (Micronerrani)	Various	Tray-loaders and -wrappers, A	30 packs

* Abbreviations: S/A Semi-automatic; A Automatic.

Case packers and tray loaders

Supplier	Machine name	Details*	Max. speed per min.
Amdelco (Systems)	Various	Tray loaders, S/A and A	40 trays
Bosch (Strunck)	Various	S/A and A	80 cases
Breda	Curlypack	Side-loading, with corrugated paper interleaving, A	300 bottles
	Vertipack	Case former, drop packer and sealer, A	20 cases
	Vertipack	Case former, packer and sealer; with descending grippers, single or multiple heads, A	20 cases per head
Conpack Engineering (Leifeld & Lemke)	Packomat	S/A and A Plus double unit	50 cases 80 cases
Crown Cork (Huntingdon)	Flowpacker	Cases raised continuously to meet moving groups of bottles, A	60 cases
(Baele)	—	A	95 cases
Creekway (Hermier)		Drop-around, S/A and A	30 cases
Emhart (Standard-Knapp)	874–12	Drop packer or descending grippers, A	48 cases

* Abbreviations: S/A Semi-automatic; A Automatic.

Case packers and tray loaders (continued)

Supplier	Machine name	Details*	Max. speed per min.
Engelmann & Buckham (Pont-a-Mousson)	Seva 72	Descending grippers, A	15 cases or trays
	Seva 101	—	25 trays
Engineering Developments	Various	Tray-loaders, S/A and A	40 trays
Erben (De Hoop) (Kettner)	Various	A	50 trays
	Various	Descending grippers, A	30 cases
Europack	TL20	A	22 trays
Fairest Trading (Certus)	Various	Descending grippers, A	67 cases
	FE1	Drop packer, A	20 crates or cases
FMC (Salwasser)	Sure-Way	S/A	25 cases
Gimson	—	S/A	160 bottles
Glenhove (Datz)	—	Descending grippers, A	41 crates
Hänsel	Traypacker	Wrap-around tray maker, A	25 trays
Hart	Wrapmatic	A	10 cases
Holstein & Kappert	—	A	85 cases
I. D. Packaging	ID.600	Tray-erector/loader and shrink-wrapper, A	600 bottles

Supplier	Machine name	Details*	Max. speed per min.
Inpac	TL	Tray-loader, A	50 trays
	TL/SA	Tray-loader, S/A	12 trays
R. A. Jones	TP Mk IV	Constant motion side-loader tray-packer	100 trays
	CMP	Constant motion side-loader case-packer	60 cases
Kane	Kanepack	Drop packers, A S/A	10 to 15 cases 6 to 8 cases
	Kanepack/Unity	Wrap-around traymaker, A	30 trays
Mather & Platt (Standard-Knapp)	874	Descending grippers, A	15 cases or trays
	Integer 855	Side-loading, A	25 cases
	879LH	Descending grippers, A	30 cases
	249	Tray-loader, A	50 trays
	202	Wrap-around tray maker, A	60 trays
	870 and 870AA	Drop-packers, A	25 cases or trays
Meadowcroft	VPI, VPII and VPIV	Vertical lowering head, A	58 crates or cases
MetaMatic (Metal Box)	Simplimatic DP-30	Single head Dual head	30 cases 60 cases
Metal Box	Kolataform	Wrap-around tray maker, S/A or A	45 trays

* Abbreviations: S/A Semi-automatic; A Automatic.

Case packers and tray loaders (continued)

Supplier	Machine name	Details*	Max. speed per min.
Meyer	Model M	A	30 cases
	Compact	A	20 cases
Miller	Airmatic	A	30 cases
Reed	Reed-Kisters Type 90	Wrap-around tray-erector/loader, A	60 trays
	Reed-Kisters Type 82	Wrap-around tray-erector/loader, A	25 trays
Remy	—	A	75 crates or cases
Rockwell/Iwema	—	A	80 trays
Stork-Amsterdam	SIH/SUH	A	60 cases
Thames Case	Pacemaker	Erector, packer, sealer for one-piece cases, S/A	—
U.D. Engineering	Mark V	A	17 crates
	Duplex	A	30 crates
Vickers-Dawson	In-line	Descending grippers S/A and A 1 to 8 heads	72 cases

* Abbreviations: S/A Semi-automatic; A Automatic.

Wrap-around case packers

Supplier	Machine name	Details*	Max. speed per min.
Emhart	EZ-1259	A	25 cases
	ACMA	A	15 cases
Engelmann & Buckham	Seva 120	A	4 cases
Hänsel	Compactomatic	A	25 cases
Inpac	PC	A	
Kane	Kanepack		—
Mather & Platt (Emhart) (Standard-Knapp)	EZ1259	A	25 cases
	Wonderwall	A; with automatic partition insertion	50 cases
Rockwell (Sunds)	Pak-Master	S/A and A; with automatic partition insertion	25 cases
Simon Packaging Systems (Focke & Pfuhl)	FPS463	Bottles packed on sides; partitions made from case blank, A	100 cases
Thames Case	Flatway	Tape or glue seal, A	—
Van den Bergh Walkinshaw (OCME)	Various	A; with automatic partition insertion	25 cases

* Abbreviations: S/A Semi-automatic; A Automatic.

Case sealers

Supplier	Machine name	Details*	Max. speed per min.
Andrew & Suter (ABC)	Various	Gluers, S/A and A	50 cases
Bosch (Strunck)	Various	Gluers and tapers, S/A and A	40 cases
Breda	Various	Gluers, tapers or combined gluers and tapers, A	30 cases
Creekway (Hermier)	Various	Gluers and tapers, S/A and A	50 cases
Emhart (ACMA)	Various	Gluers, A	20 cases
Erben (Kettner)	Various	Gluers	—
Europack	Glu-Marq	Gluers, A	15 cases
	Mini-Marq	Stapler, S/A or A	13 cases
	Marq	Stapler, A	15 cases
Fairest Trading (Certus)	KAW (Various)	Hotmelt, coldmelt glue and tape, A	32 cases
Freeze Pack	Free-lok	Tapers or gluers, S/A and A	20 cases
Glenhove (Datz)	—	Gluers, A	25 cases

Supplier	Machine name	Details*	Max. speed per min.
Kane	Various	Gluers, A	30 cases
	Tape-Rite	Taper, A	30 cases
	Centra-Line	Gluer/Taper, A	25 cases
	Kanepack/Adams Powel	Gluer/Taper, S/A and A	40 cases
	Kanepack	Hotmelt gluers, S/A and A	40 cases
Mather & Platt (Standard-Knapp)	465B	Gluer, S/A	5 cases
	476	Hotmelt gluer, A	40 cases
	455B Glu-Liner	Hotmelt gluer, A	40 cases
	Gluline Spacesaver	Gluer, A	20 cases
Minnesota (3M)	Scotch R4	Taper, A	50 cases
	Scotch TE1200 series	Tapers, S/A or A	15 cases
R. S. Platt	—	Taper, A	—
Remy	1150	Hotmelt gluer, A	100 cases
Sellotape Products	Various	Tapers, hand operated, S/A and A	—
Supertape	Various	Tapers, S/A and A	30 cases
Stork-Amsterdam	Stork-Packo	Gluer, A	40 cases
Thames Case	Pepsealer	Tapers, S/A and A	—
Wix	—	Gluers, A	40 cases
	—	Tapers, A	30 cases

* Abbreviations: S/A Semi-automatic; A Automatic.

Pallet loaders and case stackers

Supplier	Machine name	Details*	Max. speed per min.
Alvey	Series 300	Pallet loader, A	60 cases or crates
	Series 280	Pallet loader, A	25 cases or crates
	Series 310	Pallet loader, A	80 cases
Bosch (Strunck)	PBL AO1	Pallet loader, S/A	40 cases
Breda	Columbia	Interlocking crate stacker, A	60 crates
	Columbia	Pallet loaders, A	60 cases
Clayton	—	Pallet loader, S/A	15 cases
Conpack Engineering (Leifeld & Lemke)	Palamat	Pallet loader, A and S/A	85 cases
	Adoux	Case stacker, A	30 cases
Erben (Kettner)	Palopack	Pallet loader, A	—
Fairest Trading (Certus)	Various	Pallet loaders, S/A and A	47 cases
Gimson	—	Pallet loader, A	25 cases
Goddard (H. Schaefer K.G.)	Schaefer	Crate stacker, A	150 crates
		Pallet loader, A	150 cases
Holland	Various	Pallet loaders, S/A	50 cases
Holstein & Kappert	—	Pallet loader, A	85 crates
Meadowcroft	AGEB	Pallet loader, S/A and A	50 crates
Pantin	—	Pallet loader, S/A and A	80 crates or cases

Supplier	Machine name	Details*	Max. speed per min.
Rockwell/Iwema	—	Pallet loaders, S/A and A	—
Rownson	Multistak	Pallet loaders, S/A and A	—
	Trustak	Pallet loader, A	50 crates or cases
Sovex Marshall	Model G	Pallet loader, A	30 cases
	Model H	Pallet loader, A	50 cases
Stork–Amsterdam	—	Pallet loader, A	50 cases
U.D. Engineering	Duplex	Crate stacker, A	30 crates
Van den Bergh Walkinshaw	OCME	Pallet loader	50 cases
Vickers–Dawson	—	Crate stacker, A Single, Twin and Multi	52 crates

* Abbreviations: S/A Semi-automatic; A Automatic.

13. Packaging problems and how to avoid them

Many technical books include a section on 'trouble-shooting', an over-worked word meaning the action taken when things have gone wrong. It is, however, a defeatist attitude to accept troubles and rescue operations as inevitable; encouraging this attitude is the surest way of perpetuating the troubles. It is understandable that something may unexpectedly go wrong when a new process or a new machine is being operated for the first time, but this is very rarely the case in commercial packaging, and it is safe to assume that every possible packaging problem has occurred several times, and the reason for its occurrence found several times.

Therefore if troubles continue to occur, they can only be due to human failing – either failure of a supplier to provide the right materials, or failure of the user to handle the materials correctly in the light of the experience he has gained in the past, or the information and experience which is available from other people.

It is no exaggeration to say that some packers in this country and else-where have reached the stage where packaging troubles are virtually a thing of the past. The purpose of the present chapter is to give a little assistance to other packers towards this state of affairs, by passing on some of the experience of efficient packers throughout the world. Much of it is a summarised form of what has already been said in earlier chapters, so it may be regarded as a check-list of 'do's and don'ts' of packaging in glass.

Do's and don'ts

1. *Do make sure that the designer of bottles, closures and labels knows exactly what is wanted, both technically and commercially.*

Failure to mention some apparently trivial point, or failure to ask the right questions, has often caused difficulties. Even such elementary mistakes as asking for an incorrect bottle capacity, or wrong vacuity, have sometimes been made. It is of course no good asking the designer for the impossible; for example, requests have been made for 'perfectly colourless glass which cuts out all blue light'.

2. *Do give as much notice as possible of requirements to the suppliers of packaging materials.*

The packer who does this is not just being considerate to his suppliers; he is increasing his chances of receiving first-class materials at the time required. A supplier will always do his best to supply materials at short notice and 'correct to specification', but it might, for example, mean continuing to run a machine, or a furnace, when it is due for overhaul or replacement. If adequate notice is given of bottle requirements, then if unforeseen troubles* should occur during their manufacture, there will be time to remedy the situation.

3. *Do choose packaging machinery to do the job required, and do not try to alter the job (or the container design) to suit the available machinery.*

One of the main decisions to take is whether to have a 'specialised' filling line, which will handle a single type of container at very high speed and efficiency, or a general-pupose line, which will handle several jobs at moderate speeds. If a high-speed 'specialised' line is established, do not subsequently change the bottle design and expect the line to work just as it did before.

4. *Don't specify machinery requirements merely by supplying sample bottles.*

Even though some machinery manufacturers still ask for information in this form, the practice should be discouraged. Only a toleranced bottle specification provides the essential information, and unless the machinery is designed in accordance, it is a waste of time drawing up the specification, or taking care to ensure that the bottles meet the specification. If bottles or models are given to a machinery maker they should not be random samples, but should represent some definite point in the specification range, e.g. the mean or the maximum dimensions.

5. *If an automatic filling line is installed, do rationalise container designs as far as possible.*

Some packers spend a lot of money on new automatic equipment, and then try to continue to handle all the old shapes and sizes of container

---

* Bottle makers are not yet in the happy position of some of their customers, for whom production troubles never occur. It is not unknown for a glass furnace to collapse without warning.

which used to be satisfactory in their semi-automatic or manual days. This is just a waste of good machinery, and the production efficiencies achieved are usually pathetically low.

6. Do refer to the specification dimensions of bottle, closure, etc., when adjusting or setting up machinery.

It is an unsatisfactory practice to adjust the critical dimensions on a bottle-handling machine by fitting in one or two random bottles. The same remarks apply as in point 4 above.

7. Do pay all possible attention to machine maintenance.

The difference between efficiency and chaos on a filling line may be a matter of tenths of a millimetre; a badly maintained machine may soon have more than this amount of play in its bearings or other moving parts. Also, as soon as its movements start to lose their smoothness, efficiency will suffer.

8. Don't permit avoidable bottle impacts.

Some packers try to justify a badly designed filling line by saying that the bottle strength is being thoroughly tested by the impacts given. There is of course an element of truth in this – but there is much more truth in the fact that many of the bottles which apparently survive may be receiving invisible damage which will hazard their performance at a later stage.

Impacts are most damaging if received just after a glass surface has been cleaned and dried.

9. Do pay particular attention to parts of the filling line where bottles receive mechanical and thermal stresses simultaneously.

It is easy for the bottle maker to ensure that the bottles will withstand a particular mechanical or thermal stress, but when both occur together it is sometimes difficult to calculate the combined effect. For example, when bottles or jars have closures pressed on immediately after filling with a hot liquid, the two effects combine to produce considerable tension at the join of base and side wall. Even a very occasional breakage at this point will indicate that the stresses are reaching a dangerous level; a remedy would be to preheat the bottles before filling or to reduce the capping pressure.

10. *Do carry out regular checks on machine performance.*

It is not difficult to measure things such as filling heights and closure application torques, and they should be checked regularly. The inclusion of air instead of CO_2 in beverage headspaces, or too much carbonation, will increase internal pressures (not to mention its effect on quality), and periodic pressure measurements are necessary.

A feature on filling lines which often receives less attention than it should is the control of temperatures at all stages. It is sometimes thought that putting a dial thermometer on the side of a machine provides a guarantee that every bottle and its contents reach the indicated temperature, but there are three assumptions here which may be far from correct. The first is that the dial thermometer is accurate, the second is that the position where the measurement is being made is close to the bottles, and the third is that the actual glass and contents reach the temperature of the surrounding water or gas. An error of 5° for any of these reasons would make a big difference to successful bottle handling, and to product quality.

11. *Don't blame the machinery manufacturerers if trouble is encountered when running machines faster than the makers intended.*

This is not to say that machines should never be tried out at speeds faster than the makers intend; but it is a risk that may turn out to be unjustified. A better way of obtaining increased production (and increased flexibility) is to buy additional machinery.

12. *Do pay attention to bottle breakage anywhere on the line.*

Bottles do not break without a reason, and it is missing the point to remain content with $x\%$ of breakage on a filling line as being 'normal'. If there is regular breakage at any point on the line, and if the same breakage pattern is repeated, it is certain that something can be done to reduce it. (See below for instructions on fracture diagnosis.)

The above remarks apply primarily to one-trip bottles, and the situation is slightly different for multi-trip bottles. With these a small proportion may have become worn, or may have been damaged during their previous trip, and the first stresses on the filling line, usually in the washer, may be performing a useful function in eliminating any weakened bottles.

U

13. *Don't spoil the pack for a ha'porth of fibreboard.*

Remember that the case or carton is more than just a bottle holder, and has an important technical function to perform.

Fracture diagnosis

There are three occasions when a detailed examination of fractured bottles may be needed. First, as mentioned in point No.12 above, if a few breakages occur regularly at any one point in the filling line an investigation should be made to remove the cause. Secondly, if heavy breakage starts to occur at any time, the reasons must be found as quickly as possible. Finally, it sometimes happens that a bottle breaks in mysterious circumstances when in the possession of a consumer, and a fracture analysis will usually reveal what really happened.

Some fractures are extremely easy to diagnose, and any packer should be able, with a little guidance, to find the reason for them. Others are very complex, and present even the experienced glass technologist with a difficult problem. The difficulty usually arises from the speed with which breakage takes place; fractures in glass can travel at 5000 metres per second, and the stresses in the glass can change appreciably even during the few millionths of a second taken to complete the breakage. In all cases, however, the packer can and should start any investigation himself. If he can solve the matter unaided, so much the better for his production, and if he has to call in his glass supplier to help, considerable time will be saved if the packer can provide the basic information which the glass technologist will need. The instructions given below are thus not intended to provide a means of solving every breakage problem, but they represent the first steps which should be taken in every investigation.

INCIDENCE OF BREAKAGE

The first thing the glass expert will want to know is the proportion of bottles breaking, as this in itself often gives a lead to the type of defect or maladjustment responsible. If the bottles are single-trippers, or multi-trip ones being used for the first time, it is important to examine the mould numbers. Each individual bottle mould in use will have its own number, which can be found embossed on or near the base of the bottle (the numbers usually run from 1 to about 20). If several moulds are in use, and all the breakage is confined to bottles from one or two of the moulds, it suggests that there is something peculiar to bottles from these moulds. Before drawing this conclusion, it is necessary to check the mould numbers on the bottles which are not breaking, because it is quite possible that a whole consignment of bottles has come from only one or two moulds, or that the different mould numbers have become segregated into different parts of the consignment.

DIMENSIONAL CHECKS

If it is found that a small proportion of bottles are breaking, the leading dimensions of these bottles should be checked against the bottle specification, and also compared with the dimensions of the bottles which do not break. It is important that both comparisons are made, otherwise one can easily jump to the wrong conclusion. This can be shown by an example:

Suppose that the height specification of a bottle is $175 \pm 1 \cdot 3$ mm, and that some bottles break when capped. It is found that their average height is $176 \cdot 1$ mm, whereas the average height of the bottles which do not break is $174 \cdot 7$ mm. This can only mean one thing, if the height has anything to do with the breakage: that the capping machinery has been set for the wrong height.

RECONSTRUCTION AND FINDING THE FRACTURE ORIGIN

In the great majority of cases, breakage starts at a single point, and the complete breakage pattern develops as the fractures spread out from the point of origin. It is essential to find out exactly where the origin occurred by reconstructing some of the broken bottles. It is quite easy to do this if the bottles are in only two or three pieces, but it requires a little time and patience – and quite a lot of sticky tape – if the breakage is complex. In the latter case it is important to see that all the pieces of the broken bottle are retrieved, and kept separate from those of other broken bottles. Very often the smallest pieces come from nearest the origin, and so are the most important ones in the reconstruction.

When the jigsaw puzzle is complete, it may be obvious where the origin is, especially if all the fractures appear to be spreading out from one point. It helps to remember that fractures often divide into two, producing a characteristic acute-angled 'fork' as they travel through the glass. It is very rare for two separate cracks to merge together into one, so the presence of a fork is a good guide to the direction of travel. For example, in Fig. 13.1a it is almost certain that the origin must be at the point marked; in Fig. 13.1b it is equally certain that the origin cannot be at the point marked.

In less obvious cases it is necessary to examine the surface of the fractures to find the direction of travel. Fig. 13.2 shows the main diagnostic marks, some or all of which can usually be found by careful examination of a clean fractured edge.

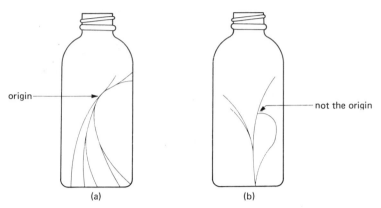

Figure 13.1.   Finding the origin of breakage

Even when a particular piece of the glass does not contain the origin, the rib marks can usually be found. The explanation of the rib marks is that they are places where the fracture has paused momentarily, before continuing at a slightly different angle; therefore they represent the leading edge of the fracture at different instants of time.

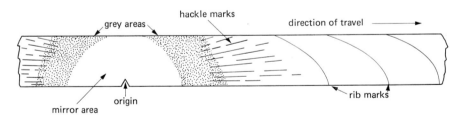

Figure 13.2.   Diagnostic markings on edges of fractured glass

The origin is always at a glass surface, and the shape of the rib marks gives a good guide as to whether the origin was on the outside or the inside surface. In Fig 13.2 the leading edge of the rib marks is on the lower surface, which will suggest an origin on that surface. (This is not by itself conclusive evidence, because occasionally the leading edge will change sides when the fracture gets further away from the origin, if it runs into an area where the stresses are very different from those at the origin.)

SOME CHARACTERISTIC BREAKAGE PATTERNS

Having found the origin, the reason for the breakage can be sought. Basically the reason is always the same: that because of applied stresses the tension in the surface at the origin has become too high. The problem is to discover what produced the tension. Thermal stresses and internal pressures often produce characteristic fracture patterns which can be easily recognised; impact stresses sometimes do this, too, but generally they are more difficult to diagnose. It is certainly not safe to conclude, from the presence of surface damage at the origin, that impact was the primary cause, nor does the absence of damage at the origin prove that impact was not responsible.

THERMAL SHOCK PATTERNS

If a bottle of simple shape is given an excessive thermal shock, for example by filling it with very hot liquid or by uniformly chilling the outside, the origin will probably be near the edge of the base, and the fracture may look as shown in Fig. 13.3a. The tension produced in this way will be on the outside surface, so the origin will also be on this surface. (If the inside surface of the bottle has been damaged or scratched, it is possible for a reverse thermal shock – chilling the inside or heating the outside – to cause thermal breakage at the inside surface.)

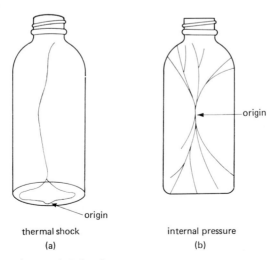

thermal shock
(a)

internal pressure
(b)

Figure 13.3.   Two characteristic breakage patterns

INTERNAL PRESSURE PATTERNS

These are the easiest patterns to recognise, taking the form shown in Fig. 13.3b.

The direction of the original fracture is always at right angles to the maximum stress, and this means that with nearly all common bottle shapes the origin of pressure breakages will be a short straight vertical line, spreading out top and bottom into a pattern of multiple forks. The number of forks is a reliable indication of the magnitude of the internal pressure at the time; with experience a glass technologist might, for example, be able to say that the bottle illustrated burst at a pressure of about 10 bars.

The origin will be on the outside surface, and the leading edges of the fractures will stay on this surface, indeed some of the smaller cracks may be on the outside surface only, not penetrating to the inside surface.

A word of warning – there is more than one way of producing a pressure breakage besides applying a steady internal pressure. For example, when a filled bottle is dropped squarely on its base the stress distribution in the glass may produce a breakage pattern exactly as illustrated in Fig. 13.3b.

IMPACT PATTERNS

When the outside surface of the body of a bottle is struck, the impact pushes the glass inwards slightly, and produces a small surrounding region of tension. This may result in a pattern of radiating cracks not unlike those of a pressure breakage; the essential difference is that the leading edge of these cracks is usually on the inside surface, and some of the shorter incomplete cracks end up running along the inside surface only.

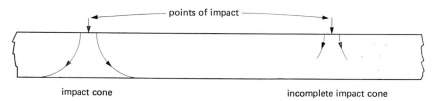

points of impact

impact cone   incomplete impact cone

Figure 13.4.   Section through impact cones

The small area in contact with the impacting object is sometimes crushed, causing a bruise or similar mark. Immediately surrounding this compressed area on the outer surface there will be a ring under tension; there is often a family of cracks starting from this ring, and spreading down into the body of the glass to form a cone shape (Fig. 13.4.). These cracks may not have time to reach the inner surface before the stress is removed, but if they do, the cone of glass may be detached in one piece, or in a number of smaller pieces.

When breakage is due to impact, there may thus be any or all of the three features:

Bruise at the point of impact.
Impact cone.
Star pattern on the inside surface.

MORE COMPLEX BREAKAGE PATTERNS

The three cases discussed so far represent the three simplest situations, and the patterns become much more difficult to sort out when more than one type of stress is being applied at the same time. The patterns are also complicated if there has been some previous damage or defect at the bottle surface; indeed it is one of the most difficult problems to decide whether a bruise or scratch at the origin of a break was produced at the moment of breakage, or whether it was there earlier, and provided a source of weakness which was subsequently 'discovered' by one of the stresses we have already discussed.

Another breakage difficult to diagnose arises from the hinge effect, i.e. when a bottle is impacted or subjected to a steady force in such a way that the maximum tension, and the origin of breakage, occurs at a place remote from the point of contact.

The complex patterns produced in these and similar cases need the attention of an expert, and an experienced bottle maker will always be ready to help.

If the reason for breakage is found to be thermal shock or internal pressure, or a combination of both, the remedial action required will usually be self-evident. But if impact or abrasion is found to be partly or entirely responsible, it sometimes requires a very careful inspection of the machinery to find where the trouble is occurring. If the impact comes from some part of the machinery itself, it can usually be prevented by dimensional adjustments or by the use of a little cushioning with rubber or plastic material. If the impact is caused by glass on glass, adjustments to bottle conveyors or guides may be needed, and in difficult cases the machinery suppliers should be consulted.

14. Glass containers and the environment

During the past few years there has been growing public concern over a wide range of ecological problems, including the dangers of environmental pollution and the need to conserve the earth's resources.

Packaging is not the only subject that has come in for criticism on both scores; indeed, there are few sectors of industrial activity that do not invite ecological comment of one sort or another. But what the critics usually forget is the positive contribution which packaging makes to the efficient distribution and preservation of the world's goods.

Certainly, problems exist – but they are not usually capable of the over-simplified solutions that are often suggested by those who are the most vociferous spokesmen on ecological issues.

What then are the environmental problems relating to packaging? Broadly they can be grouped into four main categories: (1) the litter problem; (2) pollution of the atmosphere, earth or water; (3) disfigurement of the countryside by mining, quarrying, deforestation and other methods of winning raw materials; (4) potential exhaustion of scarce resources.

In the past, these four distinct areas of debate have often been confused. The litter problem, for example, has been muddled up with the argument about returnable versus non-returnable packaging or the problems of resource conservation. Unfortunately, answers to one set of problems may not necessarily solve all the others; in fact, they may even make them worse. Let us therefore examine each of the four areas in turn in an attempt to define the problems and suggest what can or should be done about them.

I. THE THROW-AWAY SOCIETY

The casual throw-away attitude of some consumers to bottles, cans, and wrappings of all kinds is one of the main factors that has given packaging users and suppliers a flawed public image. The fact must be faced that the package which looks so right and pleasing in the supermarket or department store is an eyesore when dumped on the edge of a local beauty spot.

Even car parks and lay-bys – useful but seldom eye-catching – are made untidy by litter that could damage tyres or injure children's hands and feet.

How does glass fare in the litter stakes? In a survey commissioned by the Glass Manufacturers' Federation immediately after the 1977 Summer Bank Holiday, at fifty sites such as beaches, beauty spots, lay-bys and parks, glass packaging as a percentage of all items of litter had fallen to 1·7%, as compared with 2·3% in a similar survey undertaken in 1972. Beverage containers accounted for 7·1% in 1977 and of these the proportions were: cans 57%, glass 23%, plastics 14%, and cardboard 6%. Surprisingly, only 8% could be clearly identified as non-returnable bottles. So much for the popular theory that making all beer and soft drinks bottles returnable would solve the litter problem as far as glass containers are concerned.* It should be acknowledged, however, that the public seems sometimes uncertain whether a particular bottle is returnable or not. A need for clear identification of returnable containers is indicated here.

Despite the widespread evidence of their existence, litter-leavers are probably only an irresponsible minority of the population. Nor are they always entirely to blame. Not all local authorities provide adequate litter bins in picnic areas and lay-bys, and how many see that they are emptied often enough? Every holiday tourist knows the answer to that one.

So what is the real answer to the litter problem? The Dangerous Litter Act 1971 provides a legal sanction, though it is by no means easy to enforce. Glass manufacturers believe that the problem can best be solved by education and publicity. The Glass Manufacturers' Federation contributes to and is a founder member of the Keep Britain Tidy Group, financed jointly by industry and government. Co-operative action is therefore a fact, not a pious hope. The aim is to divert much of what is at present left as litter into the solid waste disposal or recovery system. But to do this, effective education to deter the litterbug must be combined with the provision of adequate litter-disposal facilities. This is a job not just for the glass manufacturers or the packers, but for society as a whole.

2. ENVIRONMENTAL POLLUTION

The production of glass containers involves the use of furnaces, and bottle manufacturers – like many other industrialists – are continually taking steps to curb the amount of atmospheric pollution produced by combustion processes. Glass factories produce no other emissions or effluents and all waste glass (e.g. rejects or breakages) is re-melted.

---

*It is estimated that in 1976 more than 100 million returnable soft-drink bottles went missing, representing £5 million worth of unreclaimed deposit money at 5p per bottle.

It is possible to argue that the use of returnable bottles causes pollution because of the quantitities of caustic or detergent laden water that are discharged from the washing process. In this sense, at least, non-returnable containers are preferable from an ecological point of view. In the absence of hard facts one way or the other, most people would probably agree that the potentially unwholesome effects of contaminated water released from bottle washing plants are outweighed by the benefits of an efficient returnable bottle system.

Glass itself is not a pollutant when disposed of after use; its excellent chemical inertness means that it does not degrade biologically or affect any other materials in contact with it. In terms of odour, decay and toxicity therefore, glass is entirely unobjectionable.

3. PRESERVATION OF THE COUNTRYSIDE

Spoilage of the landscape by mineral extraction processes is an unavoidable consequence of winning raw materials in an industrial society. A lot of attention is now being paid to methods of developing and landscaping quarries and sandpits in a way that will both screen the extraction and minimise its impact on the environment. Glass making, however, accounts for a very small proportion of the raw materials extracted. Only 2·4% of all the sand and gravel produced in the UK is used for making glass containers. The corresponding figure for limestone is 0·6%. If glass makers and glass users are to be castigated for their share in the spoilage of the countryside, then how much more so should builders of houses, roads, bridges – and the people who use them! Some spoilage of the landscape is unfortunately an inevitable price that we have to pay for the maintenance and improvement of our standard of living. This is not to say that we should be complacent about it. But a much greater cause for concern results from needless profligacy in the use of scarce resources, which brings us appropriately to the final and perhaps most important area of debate.

4. RESOURCE CONSERVATION

We have already noted that the raw materials which go into glass are among the most common minerals in the earth's crust. Apart from sand and limestone, which have already been mentioned, the soda ash is made from underground salt (sodium chloride) deposits, and supplies of all of these are virtually inexhaustible. And in its use of blast-furnace slag as a source of alumina, the glass maker is actually helping to use up the waste from another industry. But a glass container, like any other manufactured item, uses not only raw materials but also energy, manpower and capital – and society is becoming more and more concerned to find ways of making better use of all these resources.

There are three main ways of conserving the resources represented in a glass container. The most obvious way is to re-fill it – hence the popular call for greater use of returnable bottles, although as we shall see later this is not by any means a universal solution to the problem. The second is to break or crush the glass into fragments known as cullet, which is added to the melt in the furnace to make new bottles. The third is to use the cullet as raw material in other industries, particularly in building and construction work. All three methods of re-using glass are often labelled 'recycling', although that term is better reserved for the re-melting of cullet as distinct from the other outlets.

Discarded bottles come from three sources: the glass industry, its customers, and household waste. The industry's own waste, i.e. reject bottles and broken glass, is recycled through the melting process. Packers' waste is usually dealt with by cullet merchants who sell it to the glass and other industries, sometimes even exporting it. Only a small amount of glass in household waste is salvaged and sold to cullet merchants; the bulk is collected by local authority refuse departments and used for landfill or sent to the incinerator.

Studies have been made to discover whether a greater proportion of glass from household waste could be recycled, but usually it has been found that the cost of collection and sorting is higher than the cost of using new raw materials. So it seems unlikely that hard-pressed local authorities would be willing or able to set up sophisticated plants solely for extracting glass from domestic refuse. To be usable for re-melting, glass really does have to be clean and especially free of metal contamination including closures. (Molten iron in the bottom of a tank furnace has a dangerous drilling action, and aluminium in the batch gives rise to troublesome inclusions in the glass.) There is also the risk of upsetting a furnace if it is fed with too much strange glass of the wrong chemical composition.

In a 1971 UK survey it was found that, on average, factory cullet re-melted amounted to 19% and 'foreign' cullet to only 3% of all container glass melted. Nevertheless the glass container manufacturers have made provision for melting another 100,000 tonnes of cullet per year if it can be obtained, raising the total recycling percentage to over 25%.

The best chance of progress will arise when and if it becomes technically possible to separate *all* the reusable materials from mixed refuse, including metals which are potentially the most valuable. The development work to this end carried out by Warren Spring Laboratory is therefore of considerable interest. The other essential requirement will be to have a clear national policy on integrated recycling. For some years the glass industry has emphasised the need for this in its published studies, and now with the co-operation between government and industry that is beginning to emerge through such bodies as the Waste Management Advisory Council there is hope for the future. Also independent bodies such as INCPEN, the Industry Committee for Packaging and the Environment, representing both makers and users of packaging, have their part to play.

Apart from the cullet used in glass making, there are certain other secondary uses for waste glass. In this connection the University of Cardiff's Wolfson Laboratories are undertaking a research programme aimed at developing markets for cullet outside the glass container industry, and contact is being maintained by the university with similar investigations elsewhere.

FACTS ABOUT RETURNABLES

The idea of abandoning non-returnable beverage bottles completely and using only returnable containers has certain attractions. Closer investigation reveals that in the majority of cases, returnable bottles are not really feasible under present marketing and distribution conditions.

It might be thought that because a returnable system works well for some kinds of product and some distribution systems, it should work equally well for others. Unfortunately, this is far from being the case. In practice over 80% of all glass containers and virtually 100% of all other packaging materials are used only once for sound economic reasons.

It is well known that returnable glass bottles serve chiefly for beer, soft drinks and milk delivery requirements. The no-deposit, non-returnable bottles for beer and soft drinks were developed, like the can, largely to meet the needs of supermarkets which, being highly complex distribution chains with limited storage space, could not accept the extra cost of dealing with returned bottles.

The growth of supermarkets has greatly accelerated the trend towards nation-wide distribution of national brands of goods. (At present it is estimated that 32,000 self-service stores control 71% of retail sales, and by 1980 this will have risen to 52,000 controlling 85%.) No longer do we have a situation where manufacturers sell local products within local areas. Transport of empty bottles obviously becomes more expensive in relation to the distance travelled. The growth of non-returnable packages is thus an obvious consequence of the expansion of the supermarket movement.

Returnable systems are most successful where the delivery of full bottles can be closely integrated with the collection of empties, such as in the public house, hotel and restaurant trade – where most of the bottles do not leave the premises – and the daily household delivery system for milk.

It is, of course, possible that popular opinion (or legislation) might cause supermarkets to modify their present policy on returnable bottles. On the Continent, many supermarkets are already organised for handling returnable beer, wine and soft drink containers and the procedure has even been partly mechanised. Machines now working in Scandinavian supermarkets accept empty bottles, check them for returnable value, then issue the housewife with a credit note at the touch of a button. However, even if UK supermarkets do decide to accept returnable containers, we shall still have to persuade the public to bring them back, and this in all too many instances they fail to do, even when there is a deposit to collect.

It is interesting to recall that, in the 1963 edition of this book it was mentioned that there had recently been a swing from returnable to non-returnable bottles for proprietary wines and spirits, without any change of marketing methods or any marked change in bottle designs.

In examining the circumstances of this change it was stated that: 'At first sight the marketing methods appear just the same as for beer: most wines and spirits are consumed in licensed premises, and the empties should be readily available for collection and reuse. But distances between brewery and consumer are usually only a few miles, and the brewer runs a fleet of lorries to provide a regular delivery and collection service, whereas wine and spirit bottles are much more widely dispersed, and the empties might have to be transported several hundred miles.

'During the last decade, recovery and transport costs have apparently risen faster than the costs of new bottles, and consequently most wine and spirit bottlers have found that it has become cheaper and generally more satisfactory to use only new bottles.'

In order to see whether the situation has changed since 1963, a study for wine and spirit bottles has recently been carried out in the UK, including some practical trials. It was concluded that the best chance of success in reusing these bottles would be to set up four regional centres for sorting, washing, inspecting and transporting bottles to prospective reusers. But the cost to the packer would be about 50% greater than that of using new bottles. Evidently times have *not* changed!

ENERGY

Conservation of energy resources is becoming more and more crucial to mankind.

Visitors to glass factories cannot fail to get an impression of great heat and high energy consumption. The high temperatures (up to 1550°C) are certainly there, but in fact the glass-melting process is much less greedy of oil and/or other fuels than might be expected. This is partly because the process is a simple physical melting process, with no energy-consuming chemical reactions taking place (apart from the initial evolution of carbon dioxide), and partly because great efforts have been made during the last generation to improve furnace efficiency. Analyses by the British Glass Industry Research Association have shown that average glass furnace efficiency has recently been improving at about 4% per year.

In 1975, INCPEN studied the energy requirements for making and delivering all types of packaging material, and published their findings as follows, expressing the total energy used in equivalent tonnes of oil required to make and deliver 1 tonne of containers (TOE/tonne):

Material	Tinplate	Aluminium	Plastics	Paper	Glass
Raw material production	1·0	6·0	2·3	1·45	0·35
Conversion to containers	0·1	0·2	0·4	0·05	
Factory heating and lighting	0·04	0·08	0·16	0·07	0·02
Transport of containers to packer	0·02	0·02	0·06	0·02	0·01
Total (TOE/tonne)	1·2	6·3	2·9	1·6	0·4

In total, the British glass container industry uses a little over 0·3% of all the energy consumed in the country (see H.M. Government *Digest of Energy Statistics* for fuller details).

There are many possible ways of further reducing energy consumption in glass making, such as building bigger, more productive, furnaces, improving combustion and insulation, making bottles with a smaller weight of glass, and (to a small extent) increasing the percentage of re-melted cullet. All of these are being pursued.

Apart from the energy used in glass melting, energy is used throughout all the stages of a bottle's life. The Glass Manufacturers' Federation has commissioned exhaustive studies of energy usage throughout the packing cycle, in order to see more clearly where improvements are most needed and most achievable. Just to give one example, the twin developments of effective surface treatment and bulk palletisation of glass have permitted significant reductions in consumption of fibreboard, and in transport fuel consumption.

It would be hypocritical to pretend that these efforts were being made solely for environmental reasons, but the end results are just as valuable even when the motives are mixed.

15. Packaging in glass – the past

It is often useful to review the present state of technology or industry in the light of what has happened in the past; sometimes a historical study can reveal unsuspected reasons for things happening in the present, or can suggest new ideas for the future. With packaging in glass it is especially interesting to trace the parallel between the development of bottle making and the changes in the patterns of bottle usage, especially in relation to food and drink. Even the history of science comes into the story here and there. It might be thought that the famous scientists of the seventeeth and eighteenth centuries would not be interested in anything so common and commercial as bottles and bottling, but this is not so. We find Boyle in 1684 publishing methods of making carbonated minerals, and later Priestley, the clergyman-scientist, experimenting in a Leeds brewery. In the nineteenth century Pasteur worked out new bottling systems for Danish brewers.

The present chapter is in no way a complete history of glass packaging, but is an attempt to describe a few of the happenings which have some relevance to present-day events.

EARLY HISTORY OF GLASS CONTAINERS: THE EGYPTIANS AND ROMANS

Glass containers were first made in Egypt several thousand years ago. They were made by winding threads of molten glass around a sand shape; they were small and no doubt expensive, and were used for cosmetics, ointments and perfumes. When the art of bottle blowing was invented, somewhere about 100–200 BC, it became possible to make larger and cheaper bottles, suitable for transporting and storing wine and oil, and these containers, mainly made at Alexandria, played an important part in the highly organised economy of the Roman Empire. Soon bottles were blown in moulds, and commercial packing in glass was well on the way. Bottles were made with embossed decorations, and some even had the maker's name or an advertisement embossed on them. Another use for glass jars was for preserving or displaying cremation ashes – and some Roman jars are still doing this very job, in museums throughout Europe.

When the Roman Empire collapsed, many of their arts and technologies were lost, including the use of glass as a packing material.

THE RENAISSANCE

That glass making survived at all after the Romans was mainly due to the rise of the Moslem Empire; some of the most colourful glassware ever made has survived in the form of decorated mosque lamps from the eighth and ninth centuries AD. When the revival of Europe started in the thirteenth century, Venice became one of the main cultural and commercial centres, and the skill of the Venetian glass makers was far ahead of any others.

The Venetians of this period regarded glass solely as an artistic medium, and no effort or expense was spared in producing wonderful drinking glasses and mirrors. But they were not interested in ordinary bottles. Throughout this period, up to the seventeenth century, nearly all bottles for wine, oil, etc., were made either of leather or earthenware; the only exceptions were a few small bottles and vials used for medicine.

EARLY ENGLISH BOTTLE MAKING

In addition to the Venetian experts, there were a few glass makers scattered throughout Europe, making cruder, though possibly more useful, glassware. But it needed considerable improvements in technique and some organising genius to restore bottle making to the level it had reached in Roman times. Both needs were supplied in England in the early years of the seventeenth century. The technical advance was the use

Figure 15.1. British wine bottles: dates of manufacture, from left to right, *c.* 1650, 1713, 1757, *c* 1800

2A

of coal instead of wood for glass melting – this was made compulsory in England in 1615, when it was realised that the forests were rapidly disappearing. Coal made higher melting temperatures possible, and probably reduced melting costs. The organising genius was Sir Robert Mansell, a sailor who had risen from the rank of midshipman at the time of the Spanish Armada to become an admiral and Treasurer of the Navy. When he turned his interests to glass making, he did the job properly, and by 1623 had obtained control of all the English glass makers, with over thirty glass houses throughout the country.

Within a few years of obtaining his monopoly he apparently decided to compete with the imported earthenware bottles which were the standard wine containers. He produced a strong heavy bottle of the simplest design (see Fig. 15.1) which could be made quickly and cheaply. The success of this venture was delayed by the Civil War, since wine drinking was frowned upon by the Puritans and Roundheads.

THE RESTORATION

The Restoration was more than the return of the Monarch; it marked the return of all the more comfortable things in life, not least among them being wine. It also marked the return of glass as a successful material for

Figure 15.2.   Beer bottles washed up in 1953 from a 1720 shipwreck. (This is a photographer's reconstruction, but the bottles shown are the authentic ones washed up at Sandwich)

packaging. Within a few years the glass wine bottle had practically ousted the earthenware bottle, even though these too were by now being made in England. The glass bottles were strong, cheap (about 2d. each compared with at least 8d. for earthenware ones), and they kept wine in perfect condition. Glass bottles were not merely utilitarian; they also became fashionable, and, in order to keep up with the Jones's, every family of note had their own bottles, with their name or family crest embossed on a glass seal on the side. Samuel Pepys had his first 'sealed' bottles in 1663.

Figure 15.3. British mineral bottles of the nineteenth century

Figure 15.4.   Nineteenth century bottles of Schweppes and Crosse & Blackwell Ltd.

In this period England led the world in bottle making; by 1695 annual production was nearly three million bottles, and bottles were exported to many other countries – even to Venice. This supremacy, however, was not maintained in the eighteenth century, when many new bottle works were started in Europe, especially in Holland, to make bottles 'in the English manner'. The English bottle trade meanwhile was severely hampered by an Excise Tax (1696 to 1699 and 1745 to 1845), which was sometimes as much as a penny per bottle.* Nevertheless, progress was made in bottle design and methods of manufacture, and also in the design of bottle closures. It has been estimated that up to 1900 over 20,000 patents for new bottle closures were taken out in the UK and USA. (Some of these were a little fanciful, including a non-refillable bottle with an explosive charge in the closure, to discourage tampering.)

Brief notes on some of the happenings from 1660 up to 1939 are given in the following pages. They are set out in two columns, to illustrate how developments in bottle making went hand in hand with progress in other directions.

---

* A penny per bottle was very severe, but the tax on some filled bottles was even steeper. Early bottles of soda water carried a tax of 3d., being ranked as 'patent medicine'.

History of bottle making and usage

Bottle Making

1660. The start of the modern packaging industry – English wine bottles. The earliest shape was almost spherical, with a long neck.

c. 1740. Gradual evolution of the straight-sided wine bottle. Wines were by now being matured in bottle, and the new shape was better for storage (Fig. 15.1).

c. 1750. Several new British bottle works were being started, including some that are still operating today. Alloa Glassworks opened in 1750, and a works in Rotherham, later to become Beatson, Clark Ltd, in 1751. The most important glass-making region, however, was London.

c. 1760. It is uncertain exactly when the first wide-mouth jars were made, but they were quite common in the latter half of the eighteenth century. They were often called 'cook's bottles' and were mainly used for storing pickles.

Bottle Usage

1685. Boyle published the first ideas about making carbonated mineral waters.

c. 1700. Champagne was invented by Dom Perignon; this was only possible because strong bottles and tight-fitting corks were available.

1720. Wine bottles were also used for transporting ale. Beer bottles from a ship sunk in the North Sea in 1720 were washed up intact at Sandwich in 1953 (Fig. 15.2).

c. 1770. The scientists Cavendish and Priestley were working on the behaviour of gases dissolved in water and beer. Thomas Henry, FRS, sold the first soda water, at Manchester in 1777. He called it 'mephitic julep'.

c. 1780. Housewives had discovered how to preserve perishable foods in glass. Jane Lindesay, wife of an Irish squire, described in a letter how she cooked cocks' combs, hogs' ears or gooseberries in 'pint glass jars . . . with glass lids', and after cooking sealed the lids with carpenters' glue.

1794. Jacob Schweppe started business in Bristol as a mineral water manufacturer. He had already spent five years in partnership with others at Geneva. At first he used earthenware bottles, but they were not impermeable and he soon found that glass was necessary.

1797. Janet Keiller of Dundee sold the first orange marmalade, and so started a well-known company. Earthenware jars were used at first, but later glass.

c. 1805. An elliptical bottle with a pointed bottom was coming into use for minerals in England and France (Fig. 15.3). The only available closure at this time was the cork, which was not gas-tight unless moist, hence the need for a bottle which could not be stood on end.

1821. The bottle maker Rickett in Bristol patented a new method of making bottles in a split iron mould which determined the shape of the whole bottle including the neck. (Previously only the body of the bottle had been moulded.)

c. 1850. Bottle making was on the increase in Yorkshire, and several factories still operating today were built, such as John Lumbs of Castleford (1842) and Redfearn Bros. of Barnsley (1862).

1851. At the Great Exhibition in Hyde Park the leading exhibitor of bottles was E. Brefitt of Castleford.

1858. The Mason jar – the first domestic preserving jar with an external screw cap – was invented in the USA. The first cap was made of metal, but within a few years a glass lid was developed.

1807. During the Napoleonic wars, there was great difficulty in supplying the troops and ships with fresh foods. In 1807 the scientist Thomas Saddington published a method for preserving foods in glass, very similar to that used earlier by Jane Lindesay. The Society of Arts awarded him a prize of five guineas, but the British Government did nothing.

1810. The French chef Appert also worked out a system for food preserving, and wrote a book about it. He said: 'I chose glass as being the matter most impenetrable by air.' His method was successful, but his theory was unsound, as he thought it essential to seal the jars tightly before cooking. Fortunately most of his jars were strong enough to withstand this treatment. The French Navy made great use of the system, and Appert got a prize of 12,000 francs. After the Napoleonic wars, the London firm of Donkin & Gamble made use of Appert's method, for canned and bottled food.

1831. An Irishman revolutionised the Scotch whisky industry. Aeneas Coffey, Inspector-General of Excise in Ireland, had been inspecting distilleries so well that he was able to invent the Coffey still, which made possible large-scale distillation of grain spirit, and hence blended whisky.

1851. At the Great Exhibition over a million bottles of 'pop' were drunk.

1858. Erasmus Bond produced in England the first quinine tonic water. At about the same time ginger ale was invented in Ireland.

1863. Louis Pasteur published his famous theories, and methods for preserving food by destroying bacteria with heat. In 1870 he put his ideas into practice in a Copenhagen brewery and produced the first pasteurised beer.

1864. Donkin & Gamble (mentioned above) were taken over by Crosse & Blackwell, who were processing and packing all sorts of food (Fig. 15.4).

1869. Another shipwreck – off Australia. Bottles of Dublin Guinness brought up by divers from the hold in 1961 were found to be still drinkable, and with a 'nice head'.

1872. Two important British developments in closures were Barrett's internal screw stopper, and the Codd bottle, sealed with a captive glass marble inside the neck (Fig. 15.3). (The Codd bottle is still used today in the Far East.)
Bottle making developed in St Helens, where previously flat glass mainly had been made, when Nuttall's of Ravenhead started, followed by Cannington & Shaw in 1875.

1876. Wine drinking in Great Britain reached its peak; in this year the equivalent of 42 million bottles of wine were imported from France. (In 1975 the figure was about 80 million bottles, but the wine-drinking population was much larger.)

1880. William Ashley in Castleford invented the world's first semi-automatic bottle-making machine.

1885. Blended Scotch whisky was beginning to reach London and overseas markets. For example, James Buchanan supplied a special brand to the House of Commons, which soon became known as 'Black & White' because of its distinctive black bottle and white label.

1892. The crown cork was invented by William Painter, in USA. Americans were soon able to abandon the multiplicity of complicated closures they had been trying to use for carbonated minerals and beer.

1880 to 1897. British bottle makers lost ground to their European competitors because of prolonged labour troubles and resistance to mechanisation. But in 1897 the Woolwich firm of Moore & Nettlefolds were the first to restart making English bottles for French wine (using mainly German labour).

1886. Coca-Cola was invented in USA. Other soft drinks, such as Pepsi-Cola, followed shortly. Soda-fountains, which dispensed drinks direct from bulk supplies, were all the rage at the time, and bottled Coca-Cola did not appear until about 1896.

1903. Michael Owens built a successful fully automatic bottle making machine in Toledo. This was soon to revolutionise the world's glass industries. The first Owens machine was installed in England in 1906.

c. 1910. The mechanisation of bottle making meant the the old small bottle houses were becoming replaced by large factories with expensive machinery. Considerable capital was needed to exploit the new machines, and in 1913 five firms in St Helens, Yorkshire and Durham amalgamated to start what is now United Glass Ltd.

1920 to 1939. A further period of intensive development in closures. The first roll-on closure was made in 1923, the KNS lever cap appeared in England in 1925, and the first prise-off cap in 1926. In 1931 The Whitecap Co. introduced the 'vapour-vacuum' method of obtaining a vacuum seal, by means of steam injection just before capping.

c. 1900. Many new food products were appearing, usually in glass. For example, mayonnaise was first bottled in 1907; the first bottled pineapple juice in 1910.

1904. The first continuous tunnel pasteuriser for beer went into operation. At about the same time, automatic machinery for bottle filling and capping was beginning to appear.

1914 to 1918. During the Great War, vacuum fruit-preserving jars played an important part in maintaining British food supplies. By the end of the war 10 million jars were being made a year. The firm of Kilner Bros. became well known in this connection.

1920. Glass milk bottles had been used in small numbers, e.g. for sterilised milk, as early as 1894, but after the Great War United Dairies were the first British firm to change completely to milk delivery in glass bottles.

1926 to 1939. The start of the American baby-food industry. At first metal cans were used, but by 1939 glass was also coming into the picture.

BOTTLE COLLECTING

Thousands of the old bottles used over the last hundred years have survived intact, in museums, cellars and buried in rubbish dumps. Many of them are embossed with details of the packer and the product, so they can provide an easy and fascinating glimpse into packaging of the past. This has led to a recent rapid growth in the hobby of bottle hunting, mainly by excavation of old refuse pits. (Those who might think this sounds like dirty work can be reassured that after a few years all domestic refuse except glass and pottery disintegrates into nice clean compost.) There is now a British Bottle Collectors' Club, which assists in the exchange of information and of bottles, and organises meetings for its members.

The author, too, has indulged mildly in this hobby, and has found by digging at the bottom of his garden that his predecessors at the beginning of the century consumed (apart from beer, wine and 'pop') large quantities of Camp coffee, Eiffel Tower lemonade crystals, many brands of sauce, Scott's 'cod liver oil emulsion with lime and soda', Caley's 'eau artificielle', and, surprisingly, Saxlehner's mineral water bottled in Budapest.

WAR TIME

The period 1939–45 was, of course, a time of shortages, restrictions and every imaginable difficulty, but bottles were needed, in numbers greater than ever before, to supply the needs of the troops and the civilians. In the USA the opportunity was taken to rationalise and standardise bottle designs, with the object of producing the maximum number of strong bottles from the smallest amount of glass and fuel. This policy turned out to be of great value in the post-war years.

In the UK shortages were even more acute, and it was usually not possible to alter bottle designs at all; existing moulds had to be used till they fell to pieces. Much improvisation occurred – it was found that gin bottles made good blood-transfusion containers* – and among the very new glass-packed products coming into use were Molotov cocktails and sticky anti-tank bombs.

The post-war years will be included in the next chapter, which deals with the present and the future.

* And probably vice versa.

16. Packaging in glass — the present and future

As we saw in the last chapter, developments in glass packaging were not proceeding very rapidly in the thirties, and progress was stopped almost completely by the 1939–45 war. When, eventually, peacetime conditions returned to this country, there was a need and an opportunity for a major advance in glass packaging. This has been under way with increasing momentum since about 1950, and is likely to continue. It is logical therefore to review the immediate past and the immediate future together, to see what has been accomplished and what is likely to be achieved.

We will first consider developments in the glass industry itself, then in the techniques of bottle manufacture, and thirdly the effects of changes in products and marketing methods. Finally, developments in glass packaging techniques and container design will be reviewed.

Developments in the glass industry

SCIENCE AND TECHNOLOGY

It is well known that, sixty years ago, bottle making throughout the world was a combination of rule-of-thumb methods, instinct and a little engineering skill, but virtually no science. This was altered in 1917, when Professor W. E. S. Turner first studied the scientific problems of glass, and started in Sheffield the world's first university department to teach glass technology and carry out glass research.

Since the foundation in 1955 of a separate Research Association, the department has concentrated largely on teaching and pure research, and under the guidance of Professor Harold Rawson continues to make important contributions to glass science. Recently three separate university groups have been amalgamated into a single Department of Ceramics, Glasses and Polymers, making possible a fuller interchange of knowledge and techniques.

Scientists and technical members of the glass industry, and also of industries using glass, are able to meet regularly to exchange views, discuss problems, and often to plan co-operative research, under the auspices of the Society of Glass Technology. As a result of the increasing part played by this society in technical progress, it was found necessary some years ago to double the number of scientific journals published by the Society. These journals now contain papers and reseach reports from all over the world, and among the subjects covered there are many topics relevant to glass packaging.

INDUSTRIAL RESEARCH

Industrial glass research is an expensive business, and only a few of the largest firms are able to carry out practical research on their own account. To supplement the work of individual laboratories, and to assist firms without them, the British Glass Industry Research Association was formed in 1955. Prior to this, joint industrial research was carried out by the Department of Glass Technology at Sheffield, but the post-war expansion in technical work soon made it clear that more resources were needed. The Association's laboratories have been built in Sheffield, adjoining those of the university, and very close contact is maintained between them. The Research Association is supported financially by the great majority of glass container manufacturers, as well as by the Government.

THE GLASS MANUFACTURERS' FEDERATION

The biggest section of the GMF is concerned with the interests of British glass container manufacturers and of their customers, and through the Federation these members provide services of great value to the packaging industry. There are several technical committees dealing with everything from raw material and fuel supply to container design and standardisation; also groups involved with marketing, environmental questions, education and training, and other topics. In many cases these groups work jointly with representatives of packers' associations, suppliers of allied packaging components and machinery, and members of government and public bodies. This means that discussions between an individual customer and his supplier can, wherever necessary, be supplemented by discussions at industry level.

With our greater involvement in Europe, similar working arrangements have been established with Brussels and various overseas organisations.

Glass manufacturers, like many others, have found that it is not sufficient to rely on a single line of communication to transmit their views effectively and to obtain response to questions. Hence there is active participation in the work of bodies such as the Comité Permanent des Industries du Verre (CPIV), and the Centre Technique Internationale d'Embouteillage (CE.TIE) and the Comité Européen de Normalisation (CEN).

The GMF offices are at 19 Portland Place, London W.1 (Telephone 01–580 6952), and a call or visit on any subject to do with packaging in glass is likely to be well worth while.

Developments in bottle-making techniques

The last twenty years have been a period of intensive development in manufacturing techniques. New designs of glass furnace have been evolved, and some glass factories have been completely rebuilt, to accommodate new methods and new machinery. Throughout this work there have been two objectives: to find ways of making bottles faster, which is the surest way of making them cheaper; and to find ways of making them better, i.e. more efficient in use. There are, of course, ways of achieving one objective at the expense of the other, but this is hardly in the best interests of the user.

Bottle-making machines have been improving steadily in speed and efficiency during the last decade, partly as a result of evolutionary improvements and partly because of major changes. To give one example of the former, the sequence controls on IS machines are beginning to be converted from mechanical to electronic operation, and the greater precision obtainable should produce a small but significant improvement in the quantity and quality of bottles obtained, as well as improving the operators' working conditions.

As discussed in Chapter 2, bottles can be made much faster in multiple-cavity moulds, so many machines have been 'stretched' to permit dual manufacture of larger bottles. Also, some machines have been made bigger overall by increasing the number of machine stations from 5 or 6 to 8, 10 or even more.

Apart from the machines, much can and has been done to improve output by improving the design and/or reducing the weights of bottles. This will be reviewed in a later section.

With regard to changes leading to better bottle performance, we have already discussed the benefits of surface treatment and more sophisticated automatic inspection, and there is scope for more development here yet.

One specialised area of effort is to develop surface treatments which will remain effective throughout the indefinitely long life of a returnable bottle, not being removed by repeated exposure to hot caustic detergents. This has already become a commercial possibility for milk bottles and some carbonated minerals.

Finally, the outputs of some fast packaging operations can be materially assisted, and the unit packaging costs reduced, by the new bulk-pack systems for delivering bottles to the filling line.

DEVELOPMENT OF COMPOSITE CONTAINERS

The idea of making a container from a combination of materials is fundamentally attractive, because every material has different advantages and weaknesses and it may be possible for one material to compensate for another. (Paradoxically, closures are traditionally composite, but with modern plastics it is now possible to produce a closure from a single material.)

Combination of two materials can be done most simply by laminating flat sheets together, as with most plastics films, or by depositing one material as a thin layer on to a pre-formed sheet of the other, as with the tinning of steel plate. Unfortunately this cannot easily be done with glass containers (one can hardly count surface treatment as making a composite material, when the ratio of glass to metal oxide is about 500,000 to 1.) So other methods have to be found to make a glass–composite.

Small glass containers for aerosols with high-pressure propellants are normally coated with polyvinyl chloride (PVC) for protection and decoration by immersing heated bottles in a PVC plastisol and solidifying the PVC with further heat. The same could theoretically be done with larger bottles, but the cost would no doubt be prohibitive. However attempts have been made, mainly in Japan and the USA, to spray-coat bottles with very thin coatings of other plastics, including epoxy resins, ethyl vinyl acetate and even polyethylene. For carbonated beverage bottles, the idea is that the plastic envelope would contain all the broken pieces in the event of breakage, and that the cost of the film could be partially offset by reducing the weight of glass used.

Another very interesting approach is that of the Plastishield container, developed by Owens-Illinois in the USA for carbonated drinks and introduced in the UK in 1977 (Fig. 16.1). It consists of a lightweight bottle with an expanded polystyrene sleeve shrunk (by heat) around the body from near the finish to just inside the edge of the base. The sleeve is pre-printed or decorated as required, and the result is a container of very striking

Figure 16.1.   Plastishield bottles

appearance. The bottle can be made around 20% lighter than if unsleeved, yet it has very high resistance to impacts. In this basic form it does not provide complete restraint of glass fragments, but this could be achieved with a modified Plastishield system if really needed.

Changes in glass-packed products

New packages appear mainly when there is a new product to be marketed or when an existing product is distributed by a new method. We will consider the factors that influence such changes.

RELATIVE COST

The present competitive position of glass in relation to other packaging materials has already been discussed in Chapter 1. Looking to the future, the key facts about glass are that it is made from simple 'natural' materials, which are available in convenient locations in indefinitely large quantities. Further, there is no sign that other industries might wish to compete with the glass makers for these materials and so decrease the supply or increase the demand. Finally, there seems to be good scope for changing to other types of mineral raw material if the need should ever arise, or if the costs should make it an attractive proposition. (We already have the small example of changing from expensive synthetic alumina to crushed waste slag from steel blast-furnaces.) And as mentioned earlier there is still scope for further improvements in labour costs and fuel economy. So all in all, glass manufacturers are confident that in future the cost of glass is likely to rise less rapidly than the costs of most other packaging materials.

METHODS OF DISTRIBUTION

The marketing or distribution system can have a critical influence on the choice of container. In fact, the main reason why different packaging methods are employed in different countries is often because of the different methods of selling or distribution involved. The daily household delivery system for milk, for example, has encouraged the retention of returnable bottles in the UK to an extent unparalleled overseas. Similarly, returnable bottles are very popular containers for beer because most British beer is consumed 'on the premises' and there is a highly organised system for returning the bottles from pubs and off-licences to the brewers. Supermarkets, on the other hand, have no such system, so beers, milk and soft drinks retailed through these outlets are more often than not sold in non-returnable containers.

As glass containers are the only practical returnable packs for most ordinary retail sales, they will continue to dominate those markets where returnability is important. In other markets, packers may have a choice between paper, plastic, metal or glass containers. Let us examine this question a little more closely as far as the dairy industry is concerned.

One of the problems which has worried the dairies in recent years has been a decline in the number of trips each bottle makes from dairy to doorstep and back again. Ten years ago, most dairies sterilised and reused their bottles 40, 50 or even 80 times before they were lost or replaced. Today, the average number of trips made is probably only about 25. However, there are considerable variations in trippage rates throughout the country, and in some districts returnable milk bottles still make 40 trips or more.

The decline in trippage can be correlated with the growing sale of milk through grocery outlets and research has shown that a major proportion of the milk purchased from shops is used to 'top up' regular doorstep deliveries. One way of stemming the loss of milk bottles arising from shop sales might therefore be to improve the methods for ordering occasional extra milk from the roundsmen rather than buying casually from shops.

Even so, shops and supermarkets will continue to serve an important part of the retail milk market, and the increasing introduction by dairies of one-trip paper or plastic containers for milk sold through these outlets could be a major factor in preventing further glass bottle losses.

CONSUMER PREFERENCES

When offered a choice of pack, consumers can make their preference known by the simple process of buying one kind rather than another. There have been many examples where the force of public opinion has made itself felt in this way. One of the most spectacular of these has been in the instant coffee market. In the USA, the first instant coffee was sold in tins and glass-packed coffee did not appear until after the Second World War. The glass pack was very successful and within a few years had virtually captured the market. Later, exactly the same pattern repeated itself in the UK. The first glass jars of instant coffee appeared in England in 1960 and by 1972 they had taken almost 100% of the market from metal containers. This changeover is largely attributed to the easy opening and reseal character of the glass coffee jar.

Some manufacturers are now packing instant coffee in cartons, which although cheaper than jars, do not offer the same re-seal and storage benefits. It remains to be seen how consumers will divide their purchases between the two pack styles in the years to come.

PRODUCT DEVELOPMENT

Packaging methods are often directly related to methods of product preservation. Peas, for instance, can be dried and packed in a carton, frozen and packed in a plastic bag, or cooked and sold in a can or jar. Coffee can be sold in ground or soluble powder form, or as a bottled liquid extract. Orange juice may be sold as a frozen concentrate, a canned or bottled juice, or as a powder in a sachet.

Because of rising transport costs, there is a lot of interest in keeping the bulk of the product down, and this tends to favour dried or concentrated foods. Against the benefits of lower transport and storage costs, however, must be set the loss of quality which such processes so often entail. Many of today's research efforts in food and drink processing lie in trying to find a more satisfactory balance between economy and quality. The results may sometimes favour one kind of pack rather than another. Perhaps the most interesting new process to offer opportunities for future development as far as glass containers are concerned is aseptic packaging, described in Chapter 10.

Figure 16.2. Preserved fruits and vegetables in glass

Although everyone is on the look-out for ways of saving money, there is nearly always scope for interesting new top-quality products, even if they are a little more expensive than those already on the market. Research has shown that consumers generally associate glass-packed products with high quality, and new 'up-market' opportunities can often be tapped by packing products in glass rather than other containers which lack the same quality image.

On the Continent, for example, preserved fruits and vegetables are packed in glass to a much greater extent than in Britain, where metal cans still dominate the market. It may be said that the British are notorious for their cautious eating habits (and you can't bottle fish and chips*), but the fact remains that our shops are now selling many interesting foods in attractive glass packs from European countries on both sides of the Iron Curtain. Some British packers have already made a start with new glass-packed fruits and vegetables, and it seems likely that many others will follow in their footsteps in the future.

LEGISLATION

The main ways in which legislation can affect packaging have been outlined in Chapter 5. Among the most important future changes now in view for packers in Britain are those relating to prescribed pack sizes and tolerances under the proposed EEC 'harmonisation' programme.

Legislation also often controls the purity of products and the ways they are processed. Packaging can be a target here if there is any risk of the packaging materials affecting the purity of the contents. A short time ago, for example, PVC bottles were under test for the packaging of wines and spirits in the USA, but later the US Food and Drug Administration temporarily withdrew its approval of these bottles because of evidence that small amounts of vinyl chloride monomer could leach out into alcoholic liquids. This ruling had repercussions even on this side of the Atlantic, where it put an end to the trend towards the use of PVC flasks and miniatures by British bottlers of wines and spirits.

One conceivable topic for future legislation in Britain concerns the choice between returnable and non-returnable containers. In some parts of the world special taxes have already been imposed on non-returnable bottles for beers and soft drinks, and in Oregon, USA, ring-pull cans have been banned. We have yet to see whether or not the choice open to the packer will be similarly restricted by legislation in Britain.

LOOKING INTO THE FUTURE

To understand the reasons why different countries have different types of packaging for the same products, one may need to consider all the factors discussed above. Yet even though conditions may vary from one country to another, it is wise to keep a close watch on trends abroad, just as foreign packers need to watch what is happening here. It may be argued that changes in packaging methods are usually made, or prevented, by a

* A great pity when fish and chips are moving rapidly 'up-market'!

very small number of individuals in each country, and there is no reason why packaging executives in one country should think the same way as those in another. This may be true over a limited period of time, but a general review of the past and the present shows that if a new method is found to be commercially successful in one country, sooner or later it will appear in other countries where the commercial and social conditions are equivalent. This is not to say that what does well elsewhere is bound to succeed here, but there are occasions when valid comparisons and predictions can be made. The parallel drawn above between the USA and British markets for instant coffee is a case in point.

Figure 16.3. Wide-mouth containers with Seidel Strip closures

Developments in glass packaging techniques

We have discussed in Chapter 3 the increases in filling speeds for liquids over the last 20 years: a three-fold increase has been quite general, so beer bottles, for example, are often filled at around 800 per minute. For spirit bottles there used to be a 'speed-limit' of about 100 per minute, but now 300 is attainable. Powder filling has also been speeded up; instant coffee is packed at 400 a minute. Viscous liquids and semi-solids might be thought more difficult, but in fact modern volumetric pressure fillers can fill over 700 jam jars per minute and well over 1000 baby-food jars.

These changes have come about not as a result of completely new methods but by systematic development and expansion of conventional machines, and corresponding development of bottle designs. Although we speak generally of 'filling speeds', it is not usually the filler or the capper which needs the greatest development effort. More often than not it has been the line in-feed and the labelling operation where work has had to be concentrated.

The 'old' method of moving bottles from station to station with a plain belt conveyor has changed relatively little, although the detailed design and construction of the conveyors has improved. The author had expected to see greater use of fully synchronised lines and conveyors with individual pockets by now, but while these are very efficient when every machine on the line is running, there is an immediate loss of production for every machine stoppage. Instead, much use has been made of bottle accumulators placed before the more troublesome line components, so that bottles can queue up during breakdowns without stopping the whole line. This inevitably puts more stress on the bottles themselves, emphasising again the ever-increasing need for good functional bottle design and for good protective surface treatment.

And what of the future? It seems unlikely that there will be much need for another large increase in line speeds, and the quest is likely to be for more continuous line operation without breakdowns, and further automation to reduce labour costs.

Finally, as was said in the first edition, 'further developments in bottle closures may be expected, mainly to provide the consumer with the easiest and most efficient way of opening and reclosing the container'. A good example is the wide-mouth type of closure introduced in 1977 for carbonated drinks, discussed on page 222 (Fig. 16.3).

Developments in container design

Up to now when discussing design in this book, we have tended to concentrate on the external shape of the bottle, which has such an important effect on all its properties and characteristics. Not much has been said about choosing the optimum glass weight,* apart from remarking in Chapter 5 that the 'manufacturer will estimate the volume of glass required to meet the strength requirements'. That is a statement which could be taken literally in the past, when many of the estimates were generous enough to cover a wide range of manufacturing and handling variations. But today and in the future the determination of glass weight has to become a very exact science, because it can have such a significant effect on the speed and cost of manufacture. So there is a strong challenge to the designer to optimise the design weight while still ensuring that all the strength and performance requirements are adequately met.

There have been several well-publicised examples, and many others less well known, where bottles have been successfully 'lightweighted'

---

* Metrically speaking, we should strictly talk about 'glass mass'.

without changing the design, because the improved manufacturing techniques no longer required the original weight. In many other cases a minor or major redesign was needed.

The most striking example of both processes is provided by the pint milk bottle. The first glass bottles, of the 1920s, weighed over 20 oz. Ten years later they were 18 oz, in 1954, 14 oz, and in 1967, 12 oz. At that time a new 'dumpy' design (Fig. 16.4) was evolved and eventually proved successful in service at a weight of 250 grams (about 8·8 oz).

These changes have been possible because of the improved skills of the bottle makers and the designers, and further development for these

Figure 16.4. Ultra-lightweight milk bottle, compared with conventional design

reasons will undoubtedly continue in the future. There is, however, an additional factor which we are only just beginning to be able to allow for in the design calculations, that is the reduction in variation of the service conditions which the bottle has to withstand, and in some cases an overall reduction in the level of stresses. In the past much bottle movement was carried out by hand, often involving dropping or throwing; and some packaging machinery, too, was rather 'heavy-handed'. (A machine manufacturer once said, 'I build my machines very robustly, so that they can withstand the constant battering they have to receive from the bottles.') But with the rapid strides in mechanisation of packaging processes and elimination of some human elements, it is often found that the level of bottle stresses has been reduced, in spite of increases in operating speeds. It seems likely that with care we shall be able to allow for this increasingly in future, to the mutual advantage of bottle maker and packer.

Another direction of future trends will be towards greater rationalisation of designs for economic reasons. Glass is of course uniquely flexible in its design possibilities, and the thousands of different designs in existence show that very full advantage has been taken of this. But packers who have to handle a wide range of different bottles soon find that their filling line is not so flexible or easily adjustable, and down-time to alter conveyors and star-wheels, etc., can be very costly. So the question is, can some of the small variations in design be reduced or rationalised without losing the benefits of being able to put distinctive packs on the market? Glass designers are confident about their answer.

Returning to the visual aspects of bottle design, it used to be easy to separate designs into two groups: the simple, streamlined functional bottles, needed for high-speed and low-cost packaging; and shapes chosen for their individuality, perhaps highly decorated, for markets where costs and mechanisation were apparently not so important. The distinction between these groups is rapidly becoming blurred, for two or three reasons. Firstly, views about aesthetic design values keep changing, and at present consumers seem to be attracted to simple, flowing designs of container for both cheap and expensive products. Secondly, there are very few packers left who can afford not to be concerned with packaging costs and automation. And finally, the changes in handling and inspection methods within the glassworks have made it more difficult to produce the 'non-functional' shapes efficiently.

So there is developing a more uniform demand for containers which are at the same time light in weight, suitable for high-speed packaging operations, and capable of being recognised at point of sale as attractive packs with some individuality. The author has no doubt that glass can adequately meet this demand.

Units and conversion factors

The British system of measurement is officially based on the Système International (SI) of metric units which was established internationally in 1960. This system is built up from six fundamental units; the three which concern us here are the metre (m), the kilogram (kg) and the absolute temperature (°K). Various sub-multiples and derived units are used in practical situations, in accordance with BS 5555:1976, *SI units and recommendations for the use of their multiples and of certain other units.* This Standard also defines the spelling and abbreviations of the units.

DIMENSIONS

The metre and the millimetre (mm) are the preferred units of length, and bottle dimensions are invariably expressed in mm.

    1 inch $=25 \cdot 4$ mm (exact)
    1 inch$^2=645 \cdot 16$ mm^2 (exact)
    1 inch$^3=16{,}387$ mm^3

MASS

The kilogram and the gram (g) are the preferred units of mass; also the tonne, equal to 1000 kg, is acceptable. Strictly speaking the term 'weight' should not be used, because this means the *force* exerted by gravity on a mass but it seems unlikely that we shall ever reach the stage of saying, 'This bottle amasses 400 grams.'

    The following factors are for conversion from the avoirdupois system:

    1 grain $=0 \cdot 0648$ g
    1 dram $=1 \cdot 7718$ g
    1 ounce $=28 \cdot 3495$ g $(=16$ drams$=437 \cdot 5$ grains$)$
    1 pound $=453 \cdot 6$ g$=0 \cdot 4536$ kg
    1 ton (2240 lb) $=1 \cdot 016$ tonnes
    1 short ton (2000 lb)$=0 \cdot 907$ tonnes

FLUID CAPACITY

Even though the centimetre (cm) is frowned on as a unit of length, we have to say that the basic unit of fluid capacity, the litre (l), is equal to 1000 cm^3 and the millilitre (ml) is therefore 1 cm³. (Note that this is a welcome change since 1964, before which the millilitre was legally equal to 1·000028 cm³.)

Care is needed in converting from other capacity systems, because the old UK (imperial) units and the USA (liquid) units have the same names, but in some cases different sizes. This arose because the USA liquid gallon was equal to 0·8327 imperial gallons, and was divided into 8 pints each of 16 fl oz, compared with the imperial division into 8 pints each of 20 fl oz.

UK system		*USA system*	
1 minim	= 0·0592 ml	1 minim	= 0·06161 ml
1 fluid drachm	= 3·552 ml	1 fluid drachm	= 3·697 ml
1 fluid ounce	= 28·413 ml	1 fluid ounce	= 29·574 ml
1 pint	= 568·261 ml	1 pint	= 473·184 ml
1 gallon	= 4·5461 litres	1 gallon	= 3·7855 litres

TEMPERATURE

The practical scale of temperature is officially in degrees Celsius (°C) which is numerically identical to the more familiar degrees Centigrade. If it should ever be necessary to convert to absolute degrees Kelvin (°K), simply add 273·15.

To convert from degrees Fahrenheit (°F), subtract 32 then divide by 1·8.

PRESSURE AND STRESS

Pressure is measured as a force per unit of area, and the SI unit of force is the newton (N), a force which when applied to a mass of 1 kilogram will give it an acceleration of 1 metre/second². The derived unit most used is the meganewton per square metre (MN/m²) equal to 10^6N/m², and sometimes the kilonewton per square metre (kN/m²) equal to 10^3N/m².

For pneumatic and hydraulic pressures there is an alternative unit, the bar, which is usually more convenient. This is the pressure exerted by the earth's atmosphere at sea level under specified conditions, and is exactly equal to 0·1 MN/m², or 100 kN/m².

Pressures are often measured practically in pounds per square inch (psi), sometimes in kilograms per square centimetre, or in terms of the height of a column of water or mercury which the pressure would sustain.

$$1 \text{ psi} = 0\cdot06897 \text{ bar} = 6\cdot897 \text{ kN/m}^2$$
$$1 \text{ kg/cm}^2 = 1\cdot0138 \text{ bars} = 101\cdot38 \text{ kN/m}^2$$
$$1 \text{ metre of water} \equiv 0\cdot1014 \text{ bar} = 10\cdot14 \text{ kN/m}^2$$
$$1 \text{ metre of mercury} \equiv 1\cdot3339 \text{ bars} = 133\cdot39 \text{ kN/m}^2$$

Note that in practical use pressure in bars is usually intended to mean pressure above zero, but when discussing internal pressure in a bottle we are dealing with the *difference* between the pressures inside and outside.

Prescribed and preferred contents for pre-packed foods and drinks

In the UK many products can be offered for sale with any declared content (either by weight or volume.) Under the 1963 Weights and Measures Act such declarations are regarded as a minimum but this Act will shortly be amended, after which the declared content will be the average, thus bringing the UK into line with most other countries. This will permit the same standards of fill for pre-packaged goods for both home trade and export markets.

With some products, however, the choice of sizes is limited by legislation of which the most important is listed below, effective from the dates stated:—

1963	UK Weights and Measures Act	mandatory sizes for milk; jams, jelly preserves and marmalades; honey; treacle and syrup; coffee (including instant coffee).
1972	W. Germany – Prepackaging Order 1971, with subsequent amendments	mandatory sizes for all wines; spirits; beers; vinegar; edible (cooking) oils; milk; table waters; fruit and vegetable juices; preserved fruit and vegetables (including pickles); mayonnaise, and salad creams; mustard; and various household cleaning and polishing agents.
Jan. 1977	UK Weights and Measures (Sale of Wine) Order	mandatory sizes for wine carafes or other containers sold in restaurants (could be pre-packed).
Sep. 1977	EEC Wine Regulation 1608/76	mandatory sizes for light wines.
Jan. 1978	Canada – Consumer Packaging and Labelling Regulations	mandatory metric sizes for wines.
Jan. 1979	USA – Labelling and Advertising of Wine; Metric Standards of Fill	mandatory metric sizes for wines.
Jan. 1980	USA – Distilled Spirits Labelling and Advertising; Metric Standards of Fill	mandatory metric sizes for spirits.

In addition there is the EEC 'Liquid Foods' Directive 75/106/EEC, which provides sizes for interstate trade which must be accepted in all EEC countries, and most packers regard them also as preferred sizes for home trade as well. This Directive should have been incorporated into the national laws of France, Germany, Italy, Denmark and Luxembourg by midsummer 1976 but for UK, Eire, Holland and Belgium the option was given to defer implementation until December 1979 at the latest.

Some additional sizes are appended to the Directive as approved only until 1980, so have been omitted from the following lists, which summarise the present situation for the main categories of food and drink. (No doubt other restrictions will be added later; indeed, Directives for additional products are already under discussion. Also, an amendment to the 75/106 Directive has been proposed which could have the important effect of eliminating by 1982 the 0·35 and 0·70 litre sizes for all types of drink and also some of the smaller sizes for sparkling and carbonated drinks and fruit juices.)

Sizes are stated in litres unless otherwise specified.

CATEGORY

A	*Light Wines* (non-sparkling, produced only from grapes)	Mandatory in EEC	0·10, 0·25, 0·35, 0·375, 0·50, 0.70, 0·75, 1, 1·5, 2, 5.
		Mandatory in USA	0·10, 0·187, 0·375, 0.75, 1, 1·5, 3.
		Mandatory in Canada	0·187, 0·375, 0·50, 0·75, 1, 1·5, 2, 3, 4.
		Mandatory in UK for wine carafes	0·25, 0·50, 0·75, 1 or 10 fl oz or 20 fl oz.
B	*Other Non-Sparkling Fermented Beverages* (for example still cider, perry, mead)	Preferred in EEC	0·10, 0·25, 0·35, 0·375, 0·50, 0·70, 0·75, 1, 1·5, 2, 5.
C	*Vermouths and Liqueur Wines*	Preferred in EEC	0·10, 0·375, 0·50, 0·75, 1, 1·5.
		Mandatory in W. Germany	0·10, 0·20, 0·25, 0·375, 0·50, 0·70, 0·75, 1, 1·5, 2, 3, 5.
		Mandatory in USA	0·10, 0·187, 0·375, 0·75, 1, 1·5, 3.
		Mandatory in Canada	0·187, 0·375, 0·50, 0·75, 1, 1·5, 2, 3, 4.
D	*Sparkling Wines*	Preferred in EEC and mandatory in W. Germany	0·10, 0·125, 0·20, 0·375, 0·75, 1·5, 3.
		Mandatory in USA	0·10, 0·187, 0·375, 0·75, 1, 1·5, 3.
		Mandatory in Canada	0·187, 0·375, 0·50, 0·75, 1, 1·5, 2, 3, 4.
E	*Other Sparkling Fermented Beverages* (for example sparkling cider, perry, mead)	Preferred in EEC and mandatory in W. Germany	0·10, 0·125, 0·20, 0·375, 0·75, 1, 1·5, 3.
F	*Beer*	Preferred in EEC and mandatory in W. Germany	0·25, 0·33, 0·50, 0·75, 1, 2, 3, 4, 5.
		Mandatory in Canada	0·34, 0·625.

G	*Spirits*	Preferred in EEC and mandatory in W. Germany	0·05, 0·10, 0·20, 0·35, 0·375, 0·50, 0·70, 0·75, 1, 1·5, 2, 2·5, 3.
		Mandatory in USA	0·05, 0·20, 0·50, 0·75, 1, 1·75
H	*Vinegar and substitutes for vinegar*	Preferred in EEC and mandatory in W. Germany	0·25, 0·50, 0·75, 1, 2, 5.
I	*Edible Oils*	Preferred in EEC	0·10, 0·25, 0·50, 1, 2, 3, 5.
		Mandatory in W. Germany	0·10, 0·25, 0·375, 0·5, 0·75, 1, 2, 2·5, 3, 5.
J	*Milk*	Preferred in EEC	0·10, 0·20, 0·25, 0·50, 0·75, 1, 2, 3, 4.
		Mandatory in UK	$\frac{1}{3}$ pint, $\frac{1}{2}$ pint or multiples of $\frac{1}{2}$ pint.
		Mandatory in W. Germany	0·10, 0·20, 0·25, 0·50, 0·75, 1, 1·5, 2, 3, 4, 5.
K	*Waters* (including spa water and aerated waters)	Preferred in EEC	all volumes below 0·20, 0·20, 0·25, 0·33, 0·50, 0·70, 0·75, 1, 1·5, 2.
		Mandatory in W. Germany	0·20, 0·25, 0·33, 0·50, 0·70, 0·75, 1, 1·5, 2, 3, 4, 5.
L	*Lemonade* (including flavoured spa waters, and flavoured aerated waters)	Preferred in EEC and mandatory in W. Germany	all volumes below 0·20, 0·20, 0·25, 0·33, 0·50, 0·70, 0·75, 1, 1·5, 2, 3, 4, 5.
M	*Fruit Juices and Vegetable Juices*	Preferred in EEC	all volumes below 0·125, 0·125, 0·20, 0·25, 0·33, 0·50, 0·70, 0·75, 1, 1·5, 2, 3, 4, 5.
		Mandatory in W. Germany	0·10, 0·125, 0·20, 0·25, 0·33, 0·50, 0·70, 0·75, 1, 1·5, 2, 3, 4, 5.
N	*Jams, Marmalade, Fruit Jellies*	Preferred in EEC	(in grams) 125, 250, 500, 750, 1000, 1500, 2000, 2500, 5000, 7500, 10,000.
		Mandatory in UK	1, 2, 4, 8, 12 ounces, 1, 1·5 pounds and multiples of a pound.
		Mandatory in W. Germany	225, 450 grams.
O	*Honey*	Preferred in EEC	*(in grams) 125, 250, 500, 750, 1000, 1500, 2000, 2500, 5000, 10,000.
		Mandatory in UK	1, 2, 4, 8, 12 ounces, 1, 1·5 pounds and multiples of a pound.
P	*Instant Coffee*	Preferred in EEC	*(in grams) 50, 100, 200, 250, 500, 750, 1000.
		Mandatory in UK	1, 2, 4, 8, 12 ounces, 1, 1·5, pounds and multiples of a pound.

* These preferred EEC sizes for various products are still under negotiation in Brussels so should be regarded as provisional.

| Q | *Preserved and Semi-Preserved Fruit and Vegetables* (in brine, vinegar, syrup, etc.) | Preferred in EEC | *(these are brimful container sizes in millilitres, not the volume of contents) 106, 156, 212, 314, 370, 425, 580, 720, 850, 1062, 1700, 2550, 2650, 3100, 4250. |
| R | *Treacle and Syrup* | Mandatory in UK | 1, 2, 4, 8, 12 ounces, 1, 1·5 pounds and multiples of a pound. |

Legal controls for filled liquid contents and for Measuring Containers

The EEC Liquid Foods Directive (75/106/EEC) prescribes that if an EEC packer wishes to claim the right of assured entry into any EEC Member State for his product (indicating his claim by putting a symbolic 'e' on the label), he must control the declared average content within stated tolerances.* His packs will be subject to periodic checks by the appropriate Inspection Authority, using checking methods prescribed in the Directive. (In the UK the Inspection Authority will be the Weights and Measures Departments of the local Councils).

Checking contents

The negative tolerances permitted for contents of pre-packs are set out in Table C.I.

TABLE C.I

Nominal volume of the contents V_n(ml)	Tolerable negative error	
	% of V_n	ml
50– 100	—	4·5
100– 200	4·5	—
200– 300	—	9
300– 500	3	—
500–1000	—	15
1000–5000	1·5	

The checking of contents is based on ISO standard No. 2859 relating to methods of testing by attributes, taking a notional acceptable quality level of 2·5%. For full details of these tests the reader should consult Directive 75/106/EEC, Annex II and/or any subsequent national legislation which implements it. For ease of reference a summary of the checking method is given here.

*He must also use one of the EEC sizes specified in Appendix B.

The actual volume of the contents of the pre-packages may be measured directly by means of a volumetric measuring instrument or indirectly by weighing and measuring the product's specific gravity.

The check is carried out in two parts, the first to ensure that the individual contents are generally above the minimum permitted tolerance and the second to ensure that the average content over a batch of production meets the declared content.

Checking against minimum tolerance. The Directive permits either single or double sampling, and either destructive or non-destructive testing, the latter meaning tests in which the content is checked without opening or destroying the pack. Details are given below for the simplest option of single sampling and non-destructive testing.

The sample has to be selected from a batch representing 1 hour's production; for rates of over 500 packs/hour, the sample sizes are given in Table C.2.

TABLE C.2

Number of packs in batch	Number of packs in sample	No. of defectives	
		Accept	Reject
501–1200	80	5	6
1201–3200	125	7	8
over 3200	200	10	11

A 'defective' pack is one whose content is judged to be below the tolerable minimum, and the table shows how the sample and hence the batch is judged.

Checking the average contents. In the above cases, 50 packs are pre-selected from the main sample, their actual contents are measured, and the mean content of the 50 packs, $\bar{X}$, is calculated, also the estimated standard deviation, s. The batch is accepted if

$$\bar{X} \geqslant V_n - 0.379\ s \qquad (V_n \text{ is the nominal content.})$$

Checking Measuring Containers
(Measuring Container Directive 75/107/EEC)

Measuring Containers are similarly checked periodically at the time of manufacture, sample containers being taken from a batch representing 1 hour's production. The permitted capacity tolerances allow variation above and below nominal capacity, as follows:

TABLE C.3

Nominal capacity of measuring container V_n (ml)	Maximum permitted errors + or −	
	% of V_n	ml
50– 100	—	3
100– 200	3	—
200– 300	—	6
300– 500	2	—
500–1000	—	10
1000–5000	I	—

Testing consists of weighing the bottles empty, then again filled with water at 20°C. (In practice obviously the results could be adjusted if the temperature is not exactly 20°C, as explained in Chapter 6.) Two alternative methods of assessing the results are permitted, the 'standard deviation' method, requiring 35 containers and the 'average range' method, requiring 40 containers in the sample.

Standard Deviation Method. From the measured capacity the mean $\bar{X}$, and the estimated standard deviation, s, are calculated. The upper and lower permitted limits, T_s and T_i, are taken from Table C.3, and three non-equations have to be satisfied in order to accept the batch:

$$\bar{X} + 1.57\,s \leqslant T_s$$
$$\bar{X} - 1.57\,s \geqslant T_i$$
$$\text{and } s \leqslant 0.266\,(T_s - T_i)$$

Average Range Method. In this method the calculations are simpler, although probably longer. The 40 containers are divided chronologically into eight sub-groups of five, and the capacity range of each sub-group determined, i.e. the difference between the largest and smallest in each sub-group. The *mean range* of the eight ranges, $\bar{R}$, is then calculated, also the mean of all 40 containers, $\bar{X}$.

Again, there are three non-equations to satisfy for acceptance:

$$\bar{X} + 0.668\ \bar{R} \leqslant T_s$$
$$\bar{X} - 0.668\ \bar{R} \geqslant T_i$$
$$\bar{R} \leqslant 0.628\,(T_s - T_i)$$

Statistics and tolerances

Modern statistical theory has become a very complex subject, and those who wish to study it in detail are referred to one of the standard textbooks. This appendix contains the elements in the simplest form, and includes all that a packer needs to know to understand this book, and to apply it to practical packaging questions.

If a group of manufactured articles is selected from a large batch, and a property, say the length, is measured, there will usually be a spread in the measured values, which may be expressed in the form of a frequency distribution diagram or histogram, as in Fig. D.1.a. The spread arises partly from the random errors in the actual measurements and partly from the small deviations occurring in manufacture. If the errors in manufacture are also due to *random* factors, then it is nearly always found that the frequency distribution curve has the same general shape. (By 'random factors' we mean causes which may be present at one time and absent at another, or which may vary in magnitude in an unpredictable manner.) If a large number of repeat measurements are made on the same article, the same type of distribution curve would be obtained.

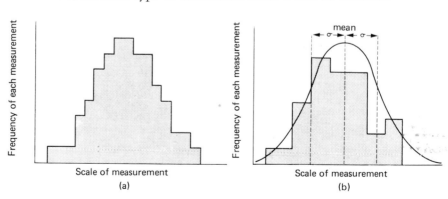

Figure D.1. Frequency distributions

STANDARD DEVIATIONS

Statistical theory can explain why the distribution curves have this characteristic shape, and more usefully it provides a method of predicting the exact shape of the curve even when only a limited number of results is available. Thus instead of all the data used to compile Fig. D.1.a we might have, say, only 30 values, which plotted by themselves as a histogram might give the unpromising result shown in Fig. D.1.b. Statistics, however, provide a simple method of calculating a quantity known as the standard deviation (often referred to as 'sigma', as the Greek letter σ is always used to represent it), and from this we can directly predict the shape of distribution curve which would be obtained if a very large number of articles was measured. Naturally, the more results we have available the more reliable will be the estimate of the shape of the curve.

From this curve (called a Gaussian curve, from the name of one of the mathematicians who discovered it over 200 years ago), we can estimate what proportion of all the articles, or of all the measurements, are likely to fall outside any limits we choose. If we choose limits at a distance equal to the standard deviation on each side of the peak, then it is predicted that 15·9% of the values will be above the upper limit, and the same percentage below the lower limit (the theoretical curve is always symmetrical).

The theoretical numbers of articles whose measured values fall outside any chosen limits are given in standard tables; a few examples are as follows:

TABLE D

Limits	Numbers falling outside each limit
$\pm \sigma$	15·9%
$\pm 2\sigma$	2·3%
$\pm 3\sigma$	0·13%
$\pm 3·5\sigma$	0·02 %
$\pm 3·75\sigma$	0·01%
$\pm 4\sigma$	0·003%
$\pm 5\sigma$	0·00003%

Practical experience has shown that $\pm 2\sigma$ is the most useful level at which to set industrial tolerances and this has been done in all the EEC packaging directives. As indicated in the table above, that is the level at which, if all the results exactly formed a Gaussian distribution, 2·3% of them would fall above the upper limit and 2·3% below the lower limit.

This, however, does not mean that in every batch or consignment there will necessarily be 23 in every 1000 outside each limit. It must be stressed that this is a theoretical figure deduced from a theoretical curve; in practice there will nearly always be reasons why the real distribution curve cannot stretch out to infinity like the theoretical one does, and of course, the supplier of the bottles will be doing his best to see that the actual curve does not stretch too far. Also,

2C

the actual curve may be slightly unsymmetrical, which will again upset the theoretical predictions.

At this stage, the reader who is new to statistics might well ask what is the use of this elaborate system if it never gives the right answer. The reply is simply that it enables us to establish a working basis for tolerances quickly and easily, and that experience has shown that decisions based on 2σ limits are usually successful in ensuring good and economic bottle performance.

MATCHING COMPONENTS

Dimensional variations are important when considering the goodness of fit of two matching components, such as a bottle neck and a closure. If the distribution curves or the standard deviations of the components are known, it is possible to calculate how well a group of closures will fit a group of bottles. Alternatively, we can calculate how much difference there should be between the mean neck diameter and the mean closure diameter.

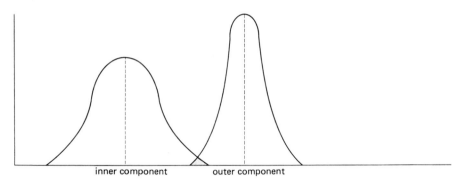

Figure D.2.   Overlapping frequency distributions

The problem can be expresssed graphically as in Fig. D.2. If the two curves represent the frequency distributions of the neck diameters and closure diameters, the area of overlap (shown shaded), expressed as a percentage of the total area under either curve, is a meaure of the percentage of times that a neck diameter is likely to be found greater than the closure diameter.

This 'percentage of overlap' may be obtained from Fig. D.3, by the following method.*

We need to know the standard deviations of the two distributions; we can, of course, deduce these if the 2-sigma limits are known and are realistic. If there must always be some clearance between the components, let the smallest

* This method, and Fig. D.3, are based on a system outlined by Hanna & Varnum in *Industrial Quality Control*, 1950, **7**, 26.

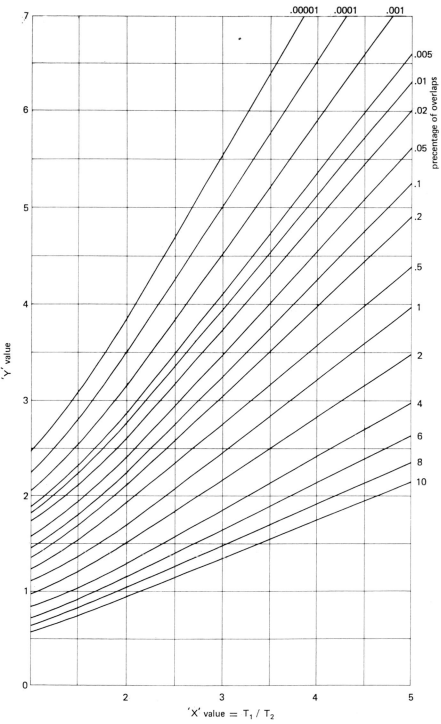

Figure D.3.

permissible clearance be C_{min}. (For a screw cap and a glass finish C_{min} is usually about 0·1 mm.) Now proceed as follows:

1. Divide the larger standard deviation by the smaller one; this is the X-value required.

2. Calculate the difference between the mean or nominal dimensions of the two components; call this D.

3. Calculate $\dfrac{D - C_{min}}{3\,s}$, where s is the *smaller* of the two standard deviations.

4. Read off from Fig. D.3 the percentage corresponding to the values of X and Y already obtained; this is the percentage of times that the dimensions will overlap.

Fig. D.3 can be used in the same way to determine the chances of some caps being loose. Suppose it is known that caps will be too loose if the clearance is greater than a certain value which we will call C_{max}.

Then the Y-value becomes $\dfrac{C_{max} - D}{3\,s}$

The X-value is $\dfrac{\text{larger } s}{\text{smaller } s}$ as before, and the graph gives the percentage of 'loose fits'.

One or two calculations of this sort should soon settle what will be the most realistic and economic specifications for a bottle and cap. These calculations can also be useful when considering the relationship between bottle tolerances and the tolerances required for various parts of the packaging machinery.

The method can be used for other properties as well as dimensions. For example, we might know that a closure for a carbonated mineral would start to leak at an excess pressure of 7 bars, with a standard deviation of 0·7 bar, and that pressures in the bottle might vary due, say, to differences in carbonation or in filling height, about a mean of 4 bars, with a standard deviation of 0·5 bar.

In this case $X = \dfrac{0·7}{0·5} = 1·4$

$C = 0$ and $D = 3$ bars

so $Y = \dfrac{3}{3 \times 0·5} = 2$

The answer obtained from Fig. D.3 is 0·02%, i.e. one bottle in 5000 may leak.

APPENDIX E

Vacuities and pressures for still* liquids

When temperature changes occur in a sealed container filled with liquid, there are three effects which cause the internal pressure to alter. Firstly, as the thermal expansion of any liquid is much greater than that of the glass, a temperature rise will make the liquid occupy a larger proportion of the space inside the container. So the gas in the headspace is compressed into a smaller volume. Secondly, a rise in temperature of the gas itself causes it to exert a higher pressure. Thirdly, there is a change in the vapour pressure exerted by the liquid; this is the pressure caused by the evaporation of some of the liquid into the air space. (In theory there is a fourth effect, the distortion of the bottle shape under pressure, but this is much too small to be significant except for bottles near their bursting point.)

Calculation of the internal pressure corresponding to any temperature and vacuity is straightforward but somewhat tedious. The equation is:

$$\frac{P_1 - S_1}{P_2 - S_2} = \frac{T_1 + 273}{T_2 + 273} \cdot \frac{[100 + V][1 + G(T_2 - T_1)] - 100 D_1/D_2}{V}$$

where T_1 is the filling temperature in °C.

T_2 is the temperature reached subsequently, in °C.

V is the percentage vacuity.

P_1 is the *absolute* pressure in the bottle when sealed at T_1.

P_2 is the *absolute* pressure in the bottle when at temperature T_2.

S_1 and S_2 are the saturated vapour pressures of the liquid at T_1 and T_2.

D_1 and D_2 are the densities of the liquid at T_1 and T_2.

G is the mean coefficient of cubical expansion of the glass ($= 25 \cdot 5 \times 10^{-6}/°C$ for bottle glass.)

The quantities P_1, P_2, S_1 and S_2 must all be measured in the same units. (The equation assumes that the air or vapour in the headspace is insoluble in the liquid, which is not exactly true, but near enough so for the present purpose.)

If, instead of the densities, the mean coefficient of cubical expansion of the liquid, L, is known, the equation may be modified by substituting

$1 + L(T_2 - T_1)$ in place of D_1/D_2.

* 'Still' means anything other than a carbonated liquid.

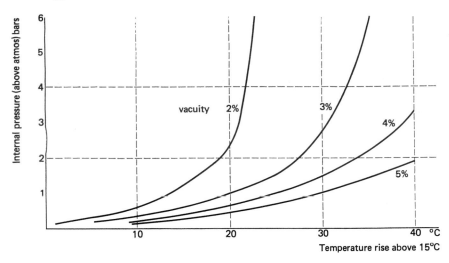

Figure E.1...Pressure rises for spirits containing 40% alcohol (70° Proof)

The thermal expansion of organic liquids, e.g. alcohols, oils, ethers, is much higher than that of water, and the pressure rise can be quite serious if the vacuity is incorrect. Fig. E.1 shows the relation between vacuity, pressure and temperature rise for an alcohol-water mixture of 70° proof strength; this is the normal strength of whisky and gin sold in the UK.

For rapid calculation of vacuities or pressure rises, the nomogram in Fig. E.2 may be used. This relates the four quantities: liquid expansion coefficient, vacuity, temperature rise and internal pressure. If two straight lines are drawn crossing scales A and B, and D and E respectively, in such a way that the two lines cut scale C at the same point, then the four values cut on scales A, B, D and E are directly related. Thus if any three of the quantities are known, the value of the fourth can be obtained directly. It will be seen from Figs. E.1 and E.2 that internal pressures can apparently rise to infinity. This represents the stage theoretically reached when the liquid has expanded sufficiently to completely fill the bottle; obviously in practice the closure would leak or the bottle would break before this happened. Water has a very variable coefficient of expansion, so it is not possible to mark a single point for it on the nomogram. If an accurate pressure rise is needed, it is better to calculate, using the known densities at T_1 and T_2. Accurate values of density, together with approximate expansion coefficients, are given in table E.

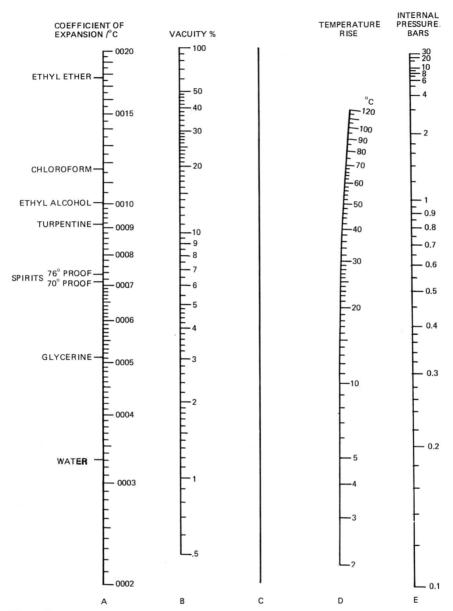

COEFFICIENT OF EXPANSION /°C	VACUITY %		TEMPERATURE RISE	INTERNAL PRESSURE. BARS

Figure E.2.

TABLE E: Density and expansion of water at various temperatures

Temperature °C	Density g/cm³	Coefficient of Expansion/°C
0	0·9999	− 0·00007
1	0·9999	− 0·00005
2	1·0000	− 0·00003
3	1·0000	− 0·00002
4	1·0000	—
5	1·0000	+ 0·00002
6	1·0000	0·00003
7	0·9999	0·00005
8	0·9999	0·00006
9	0·9998	0·00007
10	0·9997	0·00009
12	0·9995	0·00011
14	0·9993	0·00014
16	0·9990	0·00016
18	0·9986	0·00018
20	0·9982	0·00021
22	0·9978	0·00023
24	0·9973	0·00026
26	0·9968	0·00027
28	0·9963	0·00029
30	0·9957	0·00030
40	0·9922	0·00038
50	0·9881	0·00045
60	0·9832	0·00051
70	0·9778	0·00057
80	0·9718	0·00062
90	0·9653	0·00067
100	0·9584	0·00071

Vacuities and pressures for carbonated liquids

If the gas in the headspace of a sealed bottle is readily soluble in the liquid, the methods in Appendix E cannot be used. This applies particularly to carbonated beverages, where there is considerable carbon dioxide, both in the headspace and dissolved in the liquid.

The solubility of CO_2 in water decreases as the temperature increases, but it increases if the pressure is increased. As a result, when a sealed bottle of carbonated beverage is heated some of the gas which had been dissolved in the liquid comes out of solution, thus increasing the pressure in the headspace. We might expect that the smaller the vacuity the larger the pressure rise would be, but this relationship is largely obscured by the compensating effect of the gas solubility increasing with pressure. The net effect is that in a simple water–CO_2 system, the pressure is almost independent of vacuity, and only at small vacuities and high temperature is an extra effect calculable or detectable.

It is fairly straightforward to calculate approximate pressures from the usual gas laws and the published solubilities of CO_2 in water, and most tables previously published* have ignored the small effect of vacuity. The calculated data in Table F, however, published by permission of the British Glass Industry Research Association, do take vacuity into account, and the pressures have been verified by practical tests in the BGIRA Laboratories.

The carbonation, i.e. the proportion of CO_2 in the system, is given in Table F and most other sources in the usual British system of 'gas volumes': a value of two volumes of CO_2 means that two volumes of the gas (at normal temperature and pressure) are dissolved in one volume of water. The units used in other countries sometimes refer to the *weight* of gas, either as a percentage of the weight of water or in grams of gas per litre of water (1 gas volume of CO_2 is equivalent to 1·95 grams per litre or 0·195%).

Carbonation pressure tables only give equilibrium values, equilibrium being reached when there is no further evolution of gas into the headspace or solution of gas from the headspace. In a laboratory test, equilibrium can be hastened by shaking the bottle vigorously, but in many practical situations bottles are heated or cooled without movement, and in these conditions variations from equilibrium pressure may persist for some time.

---

* This includes the tables published in the first edition of this book.

There are two other reasons why the real pressures may not correspond exactly to the calculated values. The most important factor is the presence of air in the headspace or dissolved in the liquid. Oxygen and nitrogen are much less soluble in water than is CO_2, and if present they will make an additional contribution to the pressure. Air dissolved in the water will also affect the solubility of the CO_2. The effect of mixed gases in the system has not yet been calculated exactly, but experimental measurements have shown that the equilibrium values can be significantly increased, especially at low vacuity.

The second reason for deviation from theory is that the tables are based on the solubility of CO_2 in pure water. If other solids are dissolved in the water, or if the water is mixed with another liquid such as alcohol, the solubility will usually be reduced.

The situation is thus quite complicated in detail, particularly when the vacuity is low. Most carbonated beverage bottles are designed to provide vacuities of at least 3·5%, and if it is known that the contents will be reasonably air-free, then the data for 2·5% vacuity in Table F should give realistic pressure estimates. If, however, the gas composition is uncertain and the vacuity in the filled bottle less than 5%, better estimates can be obtained by taking the pressure for 5% vacuity in Table F, and increasing it by the same ratio that 5% exceeds the actual vacuity. For example, at a vacuity of 3% and a carbonation of 3 volumes, the maximum pressure at 40°C could on occasion rise as high as $^5/_3 \times 5\cdot16 = 8\cdot6$ bars above atmospheric pressure. This rule has not been worked out scientifically, but has been found by the author and others to give satisfactory guidance.

MEASUREMENT OF CARBONATION

There are exact methods available for measuring CO_2 contents by chemical or physical means, but in most situations carbonation can be estimated with sufficient accuracy by measuring the internal pressure in a sealed bottle. The procedure is simply to connect a pressure gauge to the headspace via a hollow tube or needle piercing the metal closure. Before taking a reading the bottle is shaken thoroughly to ensure that the liquid and gas are in equilibrium. The degree of carbonation can be read from Table F if the temperature of the liquid is known.

When using an ordinary Bourdon pressure gauge, the indicated pressure will be slightly lower than the true initial pressure in the bottle, unless the internal volume of the gauge is small in comparison with the volume of headspace. This can easily be allowed for by taking a first reading, P_1, then isolating the gauge from the bottle and allowing it to return to zero reading. Then reconnect with the headspace (without further shaking) and take a second reading, P_2. The initial pressure in the headspace, $P_0 = 2P_1 - P_2$.

TABLE F – Equilibrium pressures produced by various carbonations at various temperatures (CO_2 in pure water)

Vacuity %o	*Temp.* °C	*Pressure in bars – above atmospheric*			
		Carbonation – volumes			
		2	3	4	5
2·5	5	0·58	1·37	2·16	2·95
	20	1·55	2·82	4·10	5·36
	40	3·21	5·29	7·37	9·45
	60	5·32	8·38	11·46	14·52
5	5	0·58	1·37	2·16	2·95
	20	1·53	2·79	4·05	5·31
	40	3·12	5·16	7·20	9·23
	60	5·10	8·05	11·01	13·96
10	5	0·58	1·37	2·16	2·95
	20	1·49	2·73	3·97	5·21
	40	2·96	4·92	6·87	8·83
	60	4·72	7·49	10·26	13·04

Strengths of alcoholic beverages

The alcoholic strengths of beverages must be taken into account when designing bottles and closures, because it may affect the internal pressure in a filled bottle and so the vacuity and the pressure resistance may have to be considered.

Methods of measuring and expressing alcoholic strength have been in the past largely the prerogative of governments and their tax experts, and the methods used in different countries have varied widely and confusingly. Within EEC at least the situation will soon be simplified, by the adoption of a single system of expression: the percentage of alcohol by volume at 20°C. This will be measured with an EEC glass hydrometer which floats in an alcohol-water mixture. It will have a direct-reading scale in percentages of alcohol by volume.

It is not yet know exactly when this unified system will be adopted, but when it happens every EEC member will have to change from their present methods, and be able to convert from old values to new ones.

The well-known traditional method of assessing alcoholic strength was to pour the liquid over some gunpowder and try to set light to it. 'Proof spirit' was a mixture which would just burn. This method, although no doubt entertaining, lacked somewhat in precision, and the British Government first defined proof spirit in 1816 as a mixture of (ethyl) alcohol and water which at 51 °F (10·5°C) had a specific gravity of twelve-thirteenths (0·9231). This is a mixture which at 20°C contains about 58% of alcohol by volume, meaning that it could be made by diluting 58 volumes of pure alcohol with water to obtain 100 volumes of mixture.*

In the existing British system, any other alcohol–water mixture is described as so many 'degrees Proof'. The number of degrees is a measure of the number of volumes of Proof spirit which could theoretically be made from the alcohol in 100 volumes of the mixture. Thus from 100 volumes of 140° Proof spirit, 140 volumes of Proof spirit could be made by dilution, while from 100 volumes of 70° Proof spirit, 70 volumes of Proof spirit could be (indirectly) made. A common alternative way of expressing the strengths describes these two examples as 40° Over Proof (40° o.p.) and 30° Under Proof (30° u.p.). On the British scale, pure alcohol is 175·35° Proof.

* It does *not* mean that it could be made by mixing 58 volumes of alcohol with 42 volumes of water; a slight contraction occurs when the two liquids are mixed.

In the USA a similar proof system has been used, but their proof spirit is by definition a little weaker than the British one, containing exactly 50% alcohol by volume at 60°F. Therefore on the US scale pure alcohol is 200° Proof. To convert approximately from British Proof strength to US Proof strength, multiply by 1·14.

Legally acceptable tables relating the new system to the old ones and to the specific gravity of the mixtures have yet to be published; in the meantime Table G gives comparisons which should be sufficiently accurate for most purposes.

TABLE G – Alcoholic strengths

% alcohol by volume at 15·6°C	% alcohol by weight	Degrees proof (British)	Degrees proof (USA)	Specific gravity (in air) at 15·6°C
0	0·0	0·0	0	1·000
10	8·1	17·5	20	0·987
20	16·3	34·9	40	0·976
30	24·7	52·4	60	0·965
40	33·3	70·0	80	0·952
50	42·5	87·6	100	0·934
57·1	49·3	100·0	114·2	0·920
60	52·1	105·1	120	0·913
70	62·4	122·6	140	0·890
80	73·5	140·2	160	0·864
90	85·7	157·8	180	0·834
100	100·0	175·3	200	0·794

Most British beers contain up to 6% alcohol by volume, but a few very strong ones rise to about 9%. Table wines are usually in the range of 10 to 15%, and fortified wines such as port, sherry and vermouth about 20 to 21%. Home-trade spirits such as whisky and gin normally contain 40% alcohol, while 'export strength' is 43%.

Directory of suppliers

The following directory gives the addresses and telephone numbers of all the suppliers of machinery and materials, etc., mentioned in the text. Where a name has been abbreviated in the text the reader is guided to the full name by looking up the abbreviated form, e.g. against 'Pneumatic Scale' will be found '*See* Rockwell Pneumatic Scale Ltd', where the address is given. In the case of overseas manufacturers, readers are guided to the appropriate British supplier or agent.

ABC Packaging Machine Corporation. *See* Andrew & Suter Ltd

ACMA. *See* Emhart (UK) Ltd

Adams Powel Packaging Machinery Ltd. *See* Frederick C. Kane Ltd

Adelphi Manufacturing Co. Ltd, 20–21 Duncan Terrace, London, N1 8BZ (01–837 2959 and 9459)

Aerofill Ltd, Printing House Lane, Hayes, Middlesex, UB3 1AP (01–848 4501)

Albro. *See* Morgan Fairest Ltd

Alvey Conveyor Europe (UK) Ltd, 45 High Street, Egham, Surrey, TW20 9DP (Egham 6111/2)

Amdelco (Systems) Ltd, Eelmoor Road, Farnborough, Hampshire, GU14 7NW Farnborough 47171)

American Flange. *See* P. C. Bomas

Andrew & Suter Ltd, Belmont Works, Belmont Hill, St Albans, Hertfordshire, AL1 1RE (St Albans 55233/4)

Anglo Continental Machines Ltd, Marbaix House, Bessemer Road, Basingstoke, Hampshire, RG21 3NT (Basingstoke 61881)

Anker-Maschinenbau. *See* H. Erben Ltd

Arenco-Alite Ltd, Pixmore Avenue, Letchworth, Hertfordshire, SG6 1JR (Letchworth 3965/9 and 3384)

Autopack Ltd, PO Box 2, Malvern, Worcestershire, WR14 1DB (Malvern 61651)

Avery Label Systems, Gardner Road, Maidenhead, Berkshire, SL6 7PU (Maidenhead 39911)

Baele-Gangloff. *See* The Crown Cork Co. Ltd

Baele SA, G. J. *See* The Crown Cork Co. Ltd

Bausch & Ströbel. *See* Pacunion Ltd

Bancroft Fillers & Appliances Ltd, Henwood Estate, Hythe Road, Ashford, Kent, TN24 8DS (Ashford 21661/4)

Barry-Wehmiller Ltd, Bradley Fold Works, PO Box 13, Bolton, Lancashire, BL2 6RJ (Bolton 21231)

Becker Equipment & Lifts Ltd, Ealing Road, Alperton, Wembley, Middlesex, HA0 4PA (01–903 0211)

Bell Machinery. *See* Inpac Automation Ltd

Bertolaso. *See* A. Massel & Co. Ltd

Beth Label & Wrapper Machines Ltd, Henwood Estate, Hythe Road, Ashford, Kent, (Ashford 29161)

Blom & Maters Machinefabriek NV. *See* Van den Bergh Walkinshaw Ltd

Bomas, P. C., PO Box 2140 Haarlem, Holland (023 255 311)

Borden (UK) Ltd, North Baddesley, Southampton, SO5 9ZB (Southampton 732131)

Bosch Packaging Machinery (UK) Ltd, Robert, 163 Dukes Road, Acton, London, W3 0SY (01–992 5084)

Bowater Packaging Ltd, Portland House, Stag Place, London, SW1E 5DJ (01–834 9444)

Bramigk & Co. Ltd, 2a Towcester Road, Empson Street, London E3 (01–987 1051)

Breda Packaging (UK), PO Box SX14, Sittingbourne, Kent, ME9 7UF (Newington 843097)

Brevetti CEA, SAS. *See* Anglo Continental Machines Ltd

British Syphon Products (Coldflow Ltd), Hampden Park, Eastbourne, Sussex, BN22 9AJ (Eastbourne 51151)

British Miller Hydro Co. Ltd, Buckingham Avenue, Trading Estate, Slough, Buckinghamshire (Slough 23238/9 and 23230)

Brüser. *See* W. J. Hart & Sons (London) Ltd

Bunn International (UK) Ltd, B. H. Middlefield Industrial Estate, Sandy, Bedfordshire, SG19 1QW (Sandy 80851)

Busse. *See* W. J. Hart & Sons (London) Ltd

Certus Maschinenbau GmbH. *See* Fairest Trading Co. Ltd

Cioni. *See* Anglo Continental Machines Ltd

Clayton Ltd, George S., Barnaby Works, Bourne Road, Bexley, Kent (Crayford 26715 and 27913)

Cockx & Sons (London) Ltd, Windham Road, Chilton Industrial Estate, Sudbury, Suffolk (Sudbury 71176/7)

Co. Ma. Dis. *See* Anglo Continental Machines Ltd

Coster. *See* Flexile Ltd

Cozzoli Machine Company. *See* W. J. Hart & Sons (London) Ltd

Conpack Engineering Ltd. *See* Jagenberg

Creekway Ltd, SM House, 26–7 West Street, Horsham, Sussex (Horsham 61851/2/3)

Crown Cork Co. Ltd, PO Box 3, Apexes Works, Scotts Road, Southall, Middlesex (01–574 2468)

Datz. *See* Glenhove Ltd

Dawson & Barfos Manufacturing Ltd. *See* Vickers-Dawson

Dawson. *See* Vickers-Dawson

De Hoop. *See* H. Erben Ltd

Depallorator. *See* Van den Bergh Walkinshaw

Detexomat Machinery Ltd, (Packaging Division), Cressex Industrial Estate, Coronation Road, High Wycombe, Buckinghamshire (High Wycombe 21861)

Dico Packaging Engineers Ltd, Merrow, Guildford, Surrey, GU4 7BX (Guildford 60677)

Diversey Ltd, Cockfosters, Barnet, Hertfordshire (01–449 5566)

Doboy Ltd, Lyon Way, St Albans, Hertfordshire, AL4 0LE (St Albans 50667)

Doran Packaging Ltd, Heywood Industrial Estate, Heywood, Lancashire (Heywood 64944/5)

Douglas-Rownson Ltd, Daneshill West Industrial Estate, Basingstoke, Hampshire (Basingstoke 3262)

DRG Multiple Packaging Ltd, Filwood Road, Fishponds, Bristol, BS16 3SB (Bristol 656222)

Driver Southall Ltd, Tame Bridge, West Bromwich Road, Walsall (Walsall 614631–7)

Emhart AG. *See* Mather & Platt Ltd

Emhart (UK) Ltd, Crompton Road, Wheatley, Doncaster, Yorkshire (Doncaster 69101)

Engelmann & Buckham Ltd, William Curtis House, Alton, Hampshire, GU34 1HH (Alton 82421)

Engineering Developments (Farnborough) Ltd, Eelmoor Road, Farnborough, Hampshire, GU14 7NW (Farnborough 47171)

Erben Ltd, H. Hadleigh, Ipswich, IP7 6AS (Hadleigh 3011)

Europack Engineering Co. Ltd, Common Lane North, Beccles, Suffolk (Beccles 713162 and 712679)

European Machine Systems Ltd, The Agecroft Estate, Langley Road, Salford, M6 6JQ, Lancashire (061–736 8011)

F. &. D. Packaging Ltd, Greengate Street, Oldham, OL4 1DF (061–652 4321)

Fairest Trading Co. Ltd, Sorby Street, Sheffield, S4 7LB (Sheffield 27429)

Farmomac. *See* Flexile Ltd

Farrer & Son Ltd, Town Street, Rawdon, Leeds, OS19 6PU (Rawdon 502482)

Farrow & Jackson Ltd. *See* Purdy Machinery Co. Ltd

Femia. *See* PCM & Femia Ltd

Flexile Ltd, Bessemer Drive, Stevenage, Herts, SG1 2DU (Stevenage 51491)

Flower-Faraday Ltd, Eclipse Works, Wimborne, Dorset, BH21 1RE (Wimborne 2251)

FMC Corporation (UK) Ltd, Food Machinery Division, Holt Road, Fakenham, Norfolk, NR21 8JH (Fakenham 3211)

Focke & Pfuhl. *See* Simon Packaging Systems

Fords (Finsbury) Ltd, Chantry Avenue, Kempston, Bedford, MK42 7RS (Bedford 852365)

Freeze Pack Machinery Ltd, Walnut Tree Close, Guildford, Surrey, GU1 4UB (Guildford 32811)

FVR. *See* Peter Holland (Food Machinery) Ltd

General Electric (USA). *See* Landré-Mijnssen BV

Gilowy-Meteorwerk. *See* H. Erben Ltd

Gimson & Co. (Leicester) Ltd. *See* Holstein & Kappert, Gimson Ltd

Glenhove Ltd, PO Box 3, Redhill, RH1 5AY (Redhill 68725/6)

Göbel, KG Ernst. *See* M. & E. Goddard Ltd

Goddard Ltd, M. & E, Britannia Works, Sherborne Street, Manchester, M3 1EF (061–832 3161)

Grässle, Walter. *See* M. & E. Goddard Ltd

Gravfil Machines Ltd, Chapel Road, London SE27 (01–761 1481)

Hänsel Machinery Ltd, Otto, 6F Cathedral Square, Peterborough, PE1 1XH (Peterborough 60911)

Hapa. *See* Anglo Continental Machines Ltd

Harlands of Hull Ltd, Land of Green Ginger House, Anlaby, Hull, HU10 6RN (Hull 561166)

Hart & Sons (London) Ltd, W. J., 212 Putney Bridge Road, London, SW15 2NF (01–788 9295/6)

Herbort, August. *See* Mason & Morton Ltd

Hill Engineering Co. (Hull) Ltd, The Thomas, 247–9 Beverley Road, Hull, Yorkshire, HU5 2UU (Hull 43297)

Holland (Food Machinery) Ltd, Peter, St Peter's Hill, Stamford, Lincolnshire, PE9 2PE (Stamford 2667)

Holstein & Kappert, Gimson Ltd, 34 Warwick Road, Kenilworth, CV8 1HE (Kenilworth 57622/3/4)

Hunter Ltd, Thomas, Omnia Works, Rugby, CV21 1BA (Rugby 2011)

Huntingdon Industries Inc. *See* Crown Cork Co. Ltd

I. D. Packaging Ltd, Queens Works, New Hall Street, Burnley, Lancashire, BB10 1PJ (Burnley 21051)

Industrie-Werke Karlsruhe. *See* Pacunion Ltd

Inpac Automation Ltd, Uxbridge Road, Southall, Middlesex, UB1 3EZ (01–574 3673)

ITW Ltd, 470–4 Bath Road, Cippenham, Slough, Buckinghamshire (Burnham 4333)

Iwema. *See* Rockwell Pneumatic Scale Ltd
IWKA. *See* Pacunion Ltd

Jagenberg (London) Ltd, Lennig House, Masons Avenue, Croydon, CRO IEH (01–688 7423)
Johnson & Jorgensen (Plastics) Ltd, Grinstead Road, London, SE8 5AB (01–692 7551)
Jones. *See* R. A. Jones Europak Ltd
Jones Europak Ltd, R. A., 49 St Paul's Street, Leeds, LS1 2TE (Leeds 453082)
Jones, Samuel, & Co. Ltd, Butterfly House, St Neots, Huntingdon, Cambs. PE19 4EE (Huntingdon 75351)

Kane Ltd, Frederick C., Kanepack Works, Aintree Road, Perivale, Middlesex, UB6 7LA (01–998 4404)
Kemwall Engineering Co, 52 Bensham Grove, Thornton Heath, Surrey, CR4 8DA (01–653 7111)
Kettner, Max. *See* H. Erben Ltd
King, Ltd, E. C., 41 London Street, Chertsey, Surrey, KT16 8AR (Chertsey 65191–4)
Krones. *See* M. & E. Goddard Ltd
Kroneseder. *See* M. &. E. Goddard Ltd
Kugler. *See* Glenhove Ltd

Labelette Company. *See* W. J. Hart & Sons (London) Ltd
Ladewig Co., Archie. *See* Crown Cork Co. Ltd
Lamp, S. Prospero. *See* Anglo Continental Machines Ltd
Landre-Mijnssen BV, Diemen Visseringweg 40, Industrieterrein Diemen, Holland.
Leifeld & Lemke. *See* Jagenberg (London) Ltd
Lerner Machine Co. Ltd, Wharf Road, Ponders End, Enfield, Middlesex (01–804 2794)

MA Co. *See* H. Erben Ltd
Manning Ltd, A. J., Stonefield Close, South Ruislip, Middlesex, HA4 0LB (01–845 7174)
Mardon Son & Hall Ltd, Caxton House, Redcliffe Way, Bristol, BS99 7PB (Bristol 24131)
M. & M. Process Equipment Ltd, Fir Tree House, Headstone Drive, Wealdstone, Harrow, Middlesex, HA3 5QS (01–863 2282)
Massell & Co. Ltd, A., Weare Street, Ockley, Surrey, RH5 5NH (Oakwood Hill 441 and 477)
Mather & Platt Ltd, Radcliffe Works, Radcliffe, Manchester, M26 0NL (061–723 2641)
Mayflower Packaging Ltd, New Prospect Works, Meteor Close, Norwich Airport, Norwich NR6 6HQ (Norwich 416311)

Meadowcroft & Son Ltd, W., Albert Place, Lower Darwen, Blackburn BB3 0QY (Blackburn 59051)

Metal Box Ltd, Queens House, Forbury Road, Reading, RG1 3JH (Reading 581177)

Metal Closures Ltd, PO Box 32, Bromford Lane, West Bromwich, West Midlands, B70 7HY (021–553 2900)

MetaMatic (Package Handling Division of Metal Box Ltd), PO Box 3, Worcester, WR5 1HD (Worcester 54598)

Meyer Dumore International Ltd, Abbey Road, Park Royal, London, NW10 7RD (01–965 4021)

Meyer, George J. *See* Meyer Dumore International Ltd

Micronerrani S.p.A. *See* Simon Packaging Systems

Miller. *See* British Miller Hydro

Miller Hydro. *See* British Miller Hydro

Minnesota Mining & Manufacturing Co. Ltd, (3M United Kingdom Ltd), 3M House, Wigmore Street, London, W1A 1ET (01–486 5522)

Morgan Fairest Ltd, Fairway Works, Carlisle Street, Sheffield, S4 7LP (Sheffield 28751)

Nealis Harrison Ltd, Courtney Street, Hull, Yorkshire (Hull 29182)

Neri & Bonotto. *See* Anglo Continental Machines Ltd

Neumo Ltd, Quarry Road, Newhaven, Sussex, BN9 9DE (Newhaven 4301)

New Jersey. *See* Whitehall Machinery Ltd

New Way. *See* Purdy Machinery Co. Ltd

Newman Labelling Machines Ltd, Queens Road, Barnet, Hertfordshire, EN5 4DL (01–449 9666)

OCME. *See* Van den Bergh Walkinshaw Ltd

Ortmann & Herbst Maschinenfabrik. *See* H. Erben Ltd

Packaging Systems Inc. *See* Europack Engineering Co. Ltd

Pacunion Ltd, 68a Woodbridge Road, Guildford, Surrey, GU1 4RD (Guildford 73373)

Pantin Ltd, W. & C., Centre Drive, Epping, Essex, CM16 4JL (Epping 72271)

Partena S.p.A. *See* Anglo Continental Machines Ltd

PCM & Femia Ltd, 162 Windmill Road, Sunbury-on-Thames, Middlesex, TW16 7HB (Sunbury 83378)

Penrose Engineering Co. Ltd, 29 Clarence Street, Staines, Middlesex, TW18 4SY (Staines 58636/7)

Peters Packaging Ltd. *See* Whitehall Machinery Ltd

Pfaudler. *See* Peter Holland (Food Machinery) Ltd

PKB. *See* M. & E. Goddard Ltd

Phin. *See* Stackpole Machinery Company

Platt & Sons Ltd, R. S., Barkston House, Domestic Street Industrial Estate, Leeds, LS11 9RT (Leeds 36481)

Pneumatic Scale. *See* Rockwell Pneumatic Scale Ltd

Polytherm. *See* Detexomat Machinery Ltd

Pont-a-Mousson. *See* Engelmann & Buckham Ltd

Power Lifts Ltd, Hadley Works, Caxton Way, Holywell Industrial Estate, Watford, WD1 8TJ (Watford 27724)

Precision Packaging Machinery (Yorkshire) Ltd, Callywhite Lane, Dronfield, Sheffield, S18 6XR (Dronfield 6466)

Prot. *See* Anglo Continental Machines Ltd

Purdy Machinery Co. Ltd, Henwood Estate, Hythe Road, Ashford, Kent, TN24 8DS (Ashford 29161)

Raglen Packaging Ltd, Rigby Lane, Hayes, Middlesex, UB3 1ET (01–848 0541 and 0703)

Reed Medway Packaging Systems Ltd, Aylesford, Maidstone, Kent, ME20 7PQ (Maidstone 77855)

Remy & Co. Ltd, E. P., 212 Elgar Road, Reading, Berkshire, RG2 0DE (Reading 861201)

Restelli. *See* Rockwell Packaging Machines Ltd

Reynolds-Platt. *See* R. S. Platt & Sons Ltd

Roberts Filling Machines Ltd, Baldwin Street, Bolton, Lancashire, BL3 5BT (Bolton 27325)

Rockwell/Iwema. *See* Rockwell Pneumatic Scale Ltd

Rockwell Packaging Machines Ltd. *See* Rockwell Pneumatic Scale Ltd

Rockwell Pneumatic Scale Ltd, Welsh Harp, Edgware Road, London, NW2 7AA (01–452 0011)

Rose Forgrove Ltd, Seacroft, Leeds, LS14 2AN (Leeds 648177)

Rownson. *See* Douglas-Rownson Ltd

Salwasser. *See* FMC Corporation (UK) Ltd

Schaberger Carlo. *See* Rockwell Pneumatic Scale Ltd

Seitz-Werke GmbH. *See* H. Erben Ltd

Sellotape Products UK, Sellotape House, 54–8 High Street, Edgware, Middlesex, HA8 7ER (01–952 2345)

Sessions Ltd, William, The Ebor Press, York YO3 9HS (York 59224)

Simon Packaging Systems, Henry Simon Ltd, PO Box 31, Stockport, Cheshire, SK3 0RT (061–428 3600)

Simonazzi, A. & L., S. p. A. *See* Bramigk & Co. Ltd

Simplex. *See* W. J. Hart & Sons (London) Ltd

Skerman & Sons Ltd, C., 162 Windmill Road, Sunbury-on-Thames, Middlesex, TW16 7HB (Sunbury 89646)

Sovex Marshall Ltd, Carlton, Nottingham, NG4 3DY (Nottingham 249271)

SMA Pont-a-Mousson. *See* Engelmann & Buckham Ltd

Stackpole Machinery Company, 6/8 Nicholas Street, Worcester. (Worcester 20462/3)

Standard-Knapp. *See* Emhart (UK) Ltd, and Mather & Platt Ltd

Stiels Konserven-Maschinen. *See* Glenhove Ltd

Stone. *See* Engelmann & Buckham Ltd

Stork-Amsterdam International Ltd, 4 Hercies Road, Hillingdon, Middlesex, UB10 9NA (Uxbridge 51621)

Strunck. *See* Robert Bosch Packaging Machinery (UK) Ltd

Sunds AB. *See* Rockwell Pneumatic Scale Ltd

Supertape, 7 Hannington Lane, Hurstpierpoint, Sussex (Brighton 834483)

Taylor (Unity Designs) Ltd. *See* Frederick C. Kane Ltd

Thames Case Ltd, Purfleet, Essex, RM16 1RD (Purfleet 5555)

Transmatic Fyllan Ltd, Biddenham Works, Old Ford End Road, Bedford, MK40 4PF (Bedford 59296)

Udec. *See* U.D. Engineering Co. Ltd

U.D. Engineering Co. Ltd, Neills Road, St Helens, Merseyside, WA9 4TB (Marshalls Cross 813391)

U.G. Closures & Plastics Ltd, Astronaut House, Hounslow Road, Feltham, Middlesex (01–890 9051 or 5726)

Van den Bergh Walkinshaw Ltd, PO Box 19, 4 Sheet Street, Windsor, Berkshire, SL4 1BG (Windsor 51046)

Vandergeeten. *See* Vickers-Dawson

Virey-Garnier. *See* Engelmann & Buckham Ltd

Vickers Ltd, Engineering Group, Crayford Works, Crayford, Dartford, Kent, DA1 4AX (Crayford 29222)

Vickers-Dawson, Gomersal Works, Gomersal, Cleckheaton, Yorkshire, BD19 4LQ (Cleckheaton 873422)

Vickers-Vandergeeten. *See* Vickers-Dawson

Webster. *See* Vickers-Dawson

Weiss, Johann. *See* H. Erben Ltd, and Anglo Continental Machines Ltd

Whitehall Machinery Ltd, Chalks Road, Whitehall, Bristol, BS5 9ER (Bristol 553551)

Wix of London Ltd, 29/31 Minerva Road, London, NW10 6HE (01–965 1255)

World (labellers) *See* Meyer Dumore International Ltd

Zalkin & Cie Ets, Andre, *See* M. & E. Goddard Ltd

Zanasi Nigris, S.p.A. *See* Anglo Continental Machines Ltd

Societies and official bodies

British Glass Industry Research Association (BGIRA), Northumberland Road, Sheffield, s 10 2 u a (Sheffield 686201)

British Standards Institution, 2 Park Street, London w i a 2 bs (01–629 9000)

Fibreboard Packing Case Manufacturers' Association (FPCMA), 52–66 Mortimer Street, London, w i n 8 an (01–637 8451)

Glass Manufacturers' Federation, 19 Portland Place, London w i n 4 bh (01–580 6952)

Industry Committee for Packaging and the Environment (INCPEN), 39 Charing Cross Road, London, w c 2 h 0 as (01–439 3982/3)

Institute of Packaging, Fountain House, 1a Elm Park, Stanmore, Middlesex, h a 7 4 bz (01–954 6277)

National Institute for Research in Dairying, Shinfield, Reading, Berkshire (Reading 883103).

Patent Office, Southampton Buildings, London, w c 2 (01–405 8721)

Society of Glass Technology, Thornton, Hallam Gate Road, Sheffield 10 (Sheffield 663168)

Warren Spring Laboratory, Stevenage, Hertfordshire (Stevenage 3388)

Waste Management Advisory Council, Department of Industry, Millbank Tower, Millbank, London s w 1 (01–211 3940)

Index and Glossary

Brief definitions of technical terms are given below, as far as practicable. Where possible, the definitions have been taken or adapted from British Standards 3130:1974:Part 3 (Packaging Terms) and 3447:1962 (Terms used in the Glass Industry).